Climate Change and its Causes, Effects and Prediction

nova science publishers

www.novapublishers.com

Climate Change and its Causes, Effects and Prediction

Environmental Contamination and Climate Change: Effect on Plants and Remedial Strategies
Dhriti Kapoor, PhD (Editor), Renu Bhardwaj, PhD, M.Phil (Editor), Vandana Gautam, PhD (Editor)
2021. ISBN: 978-1-53619-667-2 (Hardcover)
2021. ISBN: 978-1-53619-732-7 (eBook)

Managing Climate Change
Rex S. Soto (Editor)
2021. ISBN: 978-1-53619-496-8 (Hardcover)
2021. ISBN: 978-1-53619-560-6978 (eBook)

The Caucasus and Iran: Hydrocarbons Perspectives and Impacts on the Modern World
Alexander G. Tvalchrelidze (Author), Khatuna Tabagari (Author)
2021. ISBN: 978-1-53619-382-4 (Softcover)
2021. ISBN: 978-1-53619-429-6 (eBook)

Lake Water: Properties and Uses (Case Studies of Hydrochemistry and Hydrobiology of Lakes in Northwest Russia)
Liudmila S. Shirokova (Editor), Oleg S. Pokrovsky, PhD (Editor), Taisia Y. Vorobyeva (Editor), Yulia Bespalaya (Editor)
2021. ISBN: 978-1-53619-275-9 (Hardcover)
2021. ISBN: 978-1-53619-304-6 (eBook)

Climate Change: Energy Transition, the Caribbean and Military Readiness
Carolina Parsons (Editor)
2021. ISBN: 978-1-53619-331-2 (Hardcover)
2021. ISBN: 978-1-53619-363-3 (eBook)

More information about this series can be found at
https://novapublishers.com/product-category/series/climate-change-and-its-causes-effects-and-prediction/

Satyaprakash and Anne W.M. Ng
Editors

Urban Heat Islands Reexamined

nova
science publishers
www.novapublishers.com

DOI: https://doi.org/10.52305/RKFG7202

Library of Congress Cataloging-in-Publication Data

ISBN: 979-8-88697-215-3

Published by Nova Science Publishers, Inc. † New York

Contents

Preface

Climate change is a reality, and we all are facing its effects. Urbanisation, the effect of growth, and migration of people all lead to climate change. Both these factors are leading to an increased temperature in urban areas as compared to the rural areas, where otherwise, there is no reason for the temperature to increase. This phenomenon of increased temperature in the urban areas due to dense high-rise buildings and entrapped heat is called Urban Heat Island. Scientists have been studying UHI, methods of calculation of the same and its effect on the residing population as well as flora and fauna. There have been substantiative studies on the methods of calculation of UHI, the factors leading to UHI and mitigation strategies to reduce this effect.

With the easy availability of satellite images at varied spatial resolutions and in different bands, especially the thermal band and cloud computing, the calculation of UHI has become much simpler and faster. These developments have led to many studies, and it was thought that some of these studies can be assimilated in a book for the readers to get the fundamentals of UHI, its calculation using big data and cloud computing, its effect, case studies and methods to reduce its effect on the built environment. The two editors discussed this with Nova Science Publishers and the efforts culminated in this book, Urban Heat Islands Reexamined.

We hope that the chapters of the book shall go a long way in helping the researchers, scientists and engineers in understanding UHI and its effect on the local climate and the sustainable methods which may be used to mitigate the effects of UHI and climate change.

Satyaprakash
Anne W.M. Ng

Acknowledgments

Alone we can do so little; together we can do so much - Helen Keller

Urban Heat Islands Reexamined is a depiction of the journey of various authors across the world. At the outset, the editors would like to thank each one of the authors for their contributions. Satyaprakash would like to express his gratitude to the co-editor Dr. Anne W.M. Ng, Associate Dean, College of Engineering, Information Technology and Environment, Charles Darwin University; Brinkin, Australia for reviewing and initiating the final preparation of the book. The editors would also like to thank the Nova Publishers staff for their assistance in the publication of the book, especially Nadya S. Columbus and Thomas Mazzaferro.

The editors also thankful and indebted to all others who helped in completing the book.

Chapter 1

Classification of Urbanisation and Urban Heat Island

G. S. N. V. K. S. N. Swamy, PhD
Environmental and Water Resources Engineering Division,
Department of Civil Engineering, IIT Madras, Chennai, India

Abstract

The present reflects the dynamics and conditions that built sustainable development, environmental sustainability, social equity, and climate change. Sustainable urbanization includes environmental, socio-economical, and institutional factors growth and development of city. This chapter includes the environmental factors that need to consider for urban development. The complexity of urban structure brings considerable changes to the urban atmosphere, micrometeo-rology, and dispersion conditions in a significantly considerable behavior known as urban hotspots. An increase in temperature and air pollution levels influence regional and global trends, so it is important to correlate urbanization's impact on climate change indicators. Sustainable development, mitigation, adaptation, and post-disaster planning discusses urban management and planning. This chapter aims to present the concepts mentioned in sustainable urban planning, to deal with the effects of climate change and the importance of Urban Heat Islands (UHI) in city planning. This chapter draws attention to the impact of land use and land cover change on UHI. Characterization, measurements, and impact assessment discussed in the present chapter.

Keywords: urban heat island, thermal comfort, urban air quality, secondary pollutants

In: Urban Heat Islands Reexamined
Editors: Satyaprakash and Anne W.M. Ng
ISBN: 979-8-88697-215-3

1. Introduction

To mitigate extreme events, the dynamics of climate change must be addressed as urbanization phenomena and climate variability increase in the coming years. Land use and land cover change (LULC) is one of the crucial driving factors at local, regional, and global scales of climate change. Changes in land cover are known as land-use/cover change (LUCC). Global warming is attributed to urbanization to the tune of 90%, according to the Intergovernmental Panel on Climate Change (IPCC) [1]. Through spatial analysis and numerical methods, researchers have shown that land-use conversion affects regional temperatures and precipitation as a result of urbanization [2, 3]. A principal factor in influencing regional climate was human-induced LULC. LULC has become a key topic around the world for understanding the regional climate impact. Developing countries have seen a rise in air pollution and climate change because of economic and industrial expansion. The impact of unplanned urbanization on air quality is due to several factors, including photochemical reaction rates, biogenic emissions, atmospheric circulation, and deposition. To achieve sustainable urbanization, science must apply to environmental problems from an economic and social perspective. To take corrective action on matters of environmental quality, ecological integrity, and environmental impact, two indicators identified as part of the scientific approach to sustainable urbanization. The chapter follows an introduction with four main sections. We discuss trends in urbanization and climate change in the first section. In the second section, we examine the relationship between urbanization and climate change. Global warming is being caused because of unplanned city growth and accompanied lifestyle changes and energy usage. As a result of climate change, cities have experienced changes in quality of life, competitiveness, and growth. Thirdly, the section examines urban heat islands and their impacts on local ecosystems. A variety of factors contribute to UHI. In the fourth section, we examine UHI classification and measurement techniques.

2. Urbanization Trends

Over the past decade, the migration of people from rural to urban areas has increased, 50% of the global population resides in urban areas and 60% of the world's population is expected to live in urban areas by 2020 [4]. Rapid urbanization leads to high raised buildings and surface modifications in the city areas. There are several stylized facts such as India's urban population projected to increase from 10% to 14% by 2050 and Asia's from 21 to 26% by 2050 [5]. Global, regional, and local land use and cover affect climate change. LUCC commonly use to describe this phenomenon. Global warming is attributable to urbanization, according to the IPCC's assessment report. Using spatial analysis and numerical methods, researchers have identified the impact of land-use conversion on regional temperature and precipitation due to urbanization. LULC caused by humans was one of the key factors affecting the regional climate. It has become increasingly important to understand the regional climate impact of LULC. High urban air pollution levels arise from the combination of high per capita energy consumption and large vehicular emissions in urban areas [6]. In cities, increased mortality has been linked to exposure to daytime temperatures and air pollution [7]. Additionally, haze pollution caused by changes in urban boundary air layers and atmospheric stability conditions. As a result of the haze pollution, the capacity of turbulent dispersion will be reduced [8]. In the next two decades, there are likely to be more premature deaths due to air pollution unless more aggressive targets are set and implemented. In the view of geographers, scientific advances have served as an essential propellant in the expansion of urban centers. Even though urban areas constitute only 2% of the world's land surface [9]. An urban sprawl occurs when cities rapidly extend into the countryside without any planning; in most developing countries, cities have already expanded well beyond what they normally occupy [10]. In the city landscape, urbanization has caused changes in soil distribution, rerouting of water supplies, reducing natural vegetation, adding building materials, and adding anthropogenic heat sources. Approximately 60% of urban air pollution comes from vehicular exhaust, according to literature [11, 12]. Global warming and climate change were due to urbanization, changing lifestyles, increased consumption of fossil fuels, and inefficient energy usage. However, cities also experience the effects of climate change. Cities have faced problems caused by numerous hazards associated with climate change, including rapid population growth, temperatures rising, housing and infrastructure constraints, air and water quality, and sanitation. Over the past century, the earth's average temperature has risen by roughly 0.6 degrees Celsius [13]. According to the IPCC, climate change and extreme weather events will become increasingly important in the future. From the 1990s onwards, urban climate change dynamics have been critical [14]. Adapting to the changing climate scenario requires that urban and territorial planners redefine their role and revise their skills. Urban planners have assumed that human activities plan and implementation in an "unchanging" environment, which consists of stable regional and environmental conditions. As a result of environmental change, social, economic, and environmental impacts associated with climatic phenomena, city planners are required to adopt a new perspective that increases the adaptive capacity of cities [15]. Especially in terms of mitigation and adaptation, urbanization and climate change are complex and challenging to the planners. Urban Heat Islands (UHI) are most obvious, and urban and territorial planning will need to address them structurally in the coming years [16].

3. Sustainable Urbanization Practices

When cities and territories built, environmental conditions change completely. A sustainable urban planning developed in the early twentieth century, related to hygienic needs, thermal comfort, clean water, clean air, decent housing, parks, open spaces, efficient transport systems and social welfare. Urban planning expanded to meet the emerging challenges of environmental protection, sustainable urban development and international cooperation [17]. Sustainable planning or territorial dynamics is not easy and requires attention; need to consider the consensus that exists with respect to the concept of "sustainable urbanization". Nowadays, cities are facing a new crisis, which therefore requires the strategies and principles for sustainable urban development policies. International scientific community recognizes sustainability as a major challenge for the development of the twenty-first century. Since urban areas have become central to the international debate on climate change issues, it requires an integrated and dynamic approach. Urbanization in fact identifies urban areas as a key element in the processes of globalization and transition to new land occupation models worldwide. Adapting to urbanization involves policies and practices that should be anticipatory and flexible that offset the adverse effects of climate change. City planners must include policies in the overall long-term city plans and must address the current as well as potential future impacts of climate change, climate change impacts are yet to be felt. Since policies are likely to be preventative in nature, based on strong scientific assumptions, must be tested against various scenarios. Everyone in Cities need to spread awareness and educate the public on climate change on water resources, land use, sea levels, health, and the overall ecosystems. While the adaptation policies of sustainable urbanization address adverse effects of climate change, cities need to focus on policies for mitigation to reduce their impact on the environment and increasing the green cover. As reported in many studies, emission of GHGs from cities is one of the major causes of global warming, mitigation policies must focus on reducing both direct and indirect emissions. At urban zones, the demand transportation may result in an increase in private car ownership leading to higher incidences of GHG emissions. Mitigation policies should therefore focus on reducing demand for private cars and making the technology to make cars more environmentally friendly. Policymakers need to identify the entire source of GHGs, and pollution activities supply chain, and then decide which one is the most appropriate for intervention.

The broad approach to sustainable urban development advocated is not easy to implement at any scale. This has addressed specific issues, often a particular scale. The processes of planning, as mentioned above, involve several stakeholders, and the planning arenas include a wide range of formal and informal fora. The framework for planning policies by wider policy priorities that sets international, national, regional, and local levels. Assessing the sustainable urban development by professional groups, suggesting that developers, planners, designers, the construction industry and building managers should all approach this task in the following broad sequence of actions:

- Preliminary activities
- planning of assessment activity
- what to do in assessing
- carrying out the environmental assessment
- carrying out consultations
- considering the environmental report and consultations
- providing information on the decisions
- monitoring.

Hatfield Dodds developed sustainability indicators and identified five different 'approaches' used to build support [18]. Sustainable income, maintaining ecological integrity, inequality, institutions and environmental impact, participation and sustainable well-being, alternative ethical approaches are the five approaches to achieving sustainability in urban development practice.

PICABUE analysis condenses the complex, lengthy goals of the Aalborg Charter into a shorthand form. It lays its emphasis on the need for:

- Ecological integrity
- Equity
- Public participation
- Futurity.

The most significant urban development has been the publication of the European Spatial Development Perspective (ESDP) in 1999. The ESDP has an indicative integrated policy framework for better cooperation between global, regional, states, and cities. The ESDP promotes a framework for integrated spatial policy aims to achieve the following objectives:

- A balanced and polycentric city system and a new urban-rural relationship,
- Parity of access to infrastructure and knowledge,
- Prudent management and development of the natural and cultural heritage.

4. Impact of Urbanization on UHI

Extreme temperatures in urban hotspots are becoming more common worldwide. Although a heat wave is not a recent phenomenon, the frequency and/or severity of this extreme weather event is likely to increase with the changing climate. Due to the rapid and unplanned increase in urban populations, cities are a facing an increase in pressure on their resources, infrastructure, and the environment. Water stress, sanitation, growth of slums, poor housing policies, and environmental degradation are few of the growing concerns and problems that cities are experiencing. Growth of cities affects each other and contributes towards the changing climate and global warming seriously affects city life, health, and the infrastructure.

The rapid urbanization and increasing population have necessitated the development of multistory, high-rise buildings that have altered the surface of cities. The urban heat island effect (UHI) occurs when the exhaust from motor vehicles is higher than in suburban or rural areas. Temperatures are becoming increasingly extreme in urban hotspots around the world. According to a

Canadian study, the number of days that their city's average temperature went over 30°C over the past decade had increased dramatically [19, 20]. As UHI intensity increases, indoor thermal comfort needs increase, energy use increases and air quality further deteriorate. An essential parameter for estimating local air quality is the impact of land cover change. Scientists must study the interaction between complex urban buildings, trapped heat waves and disturbed wind velocities at a height below the median net height of buildings in the downtown area. In urban environments, pavement materials have a significant impact on the daytime temperature differential. During the daytime, paved surfaces with high albedo reduce the intensity of UHI, but the effect is not as significant at night [21]. As a result, urban temperature profiles influenced not only by natural heating systems, but also by heat emissions from anthropogenic sources, such as transportation, cooking, air conditioning, and other household activities. The complex factors that impact UHI have analyzed in CFD studies to represent the diverse phenomena. Solar energy balances, ventilation of urban areas, urban heat island effects as well as pollutant transformations considered. For complete understanding of urban air quality and associated health risks under changing climatic conditions in India, accurate and reliable measurements of NOx, HC, ozone (O_3), and ultrafine PM concentrations are essential.

As in developed countries, the current Indigenous urban system tends to repeat regional urban primacy situations. Between 1951 and 2011, each city's population grew by more than 50% annually. By studying the urbanization consequences at the regional level, suggested that the local conditions for urban supremacy were gradually reabsorbed. In the 19th century [22], first reported effects of Urban Heat Island (UHI), a temperature differential phenomenon specific to cities. A British climatologist, Luke Howard for the first time studied inadvertent urban climate modifications due to various impacts.

In comparison with rural areas, urban areas rose about 3.7°C more during the nighttime than rural areas. Rural areas experienced slightly cooler temperatures during the day, about 0.34°C lower. Numerous studies have tried to explain the UHI phenomenon offered a definition with a broader perspective [23-25]. As compared to surrounding areas, these studies found that UHI had higher temperatures, higher energy demands, and higher pollution levels.

Due to the changes in the physical characteristics of the landscape brought about by urbanization, heat, mass, and momentum exchange between the surface and the atmosphere affected. Inadvertently, this resulted in a UHI effect because of an altered energy exchange. As a result of urbanization, air temperatures have undoubtedly increased in cities. A study conducted in Fujibe showed that when the population increased, the average temperature climbed [26]. There is a distinct localized climate in cities because of these artificial alterations [27]. By 2025, expected that 60-70 percent of the total world population will be living in urban areas, up from 47 percent in 2005. This urbanization progress will be obvious soon, in the coming ten years [9]. The world's most important challenge may be managing urban growth in developing countries.

5. Urban Heat Island

It is reported that in Canada, the number of days in which the average temperature exceeds 30°C has increased dramatically in recent years, due to surface modifications and massive exhaust emissions from vehicles in urban areas. This phenomenon is called the urban heat island effect [20, 28]. UHI intensification affects outdoor comfort, energy usage, and air quality. Local air quality estimations are highly dependent on land cover change. As a result of the complex structure of the urban building, heat waves and wind velocity can often trap at a height below the average building height in downtown areas. Daytime temperature differences in urban environments significantly influenced by the albedo of pavement materials in the downtown area. Urban pavements that have a high albedo reduce UHI intensity during the day, but they affect night-time temperatures significantly.

Temperature levels, energy consumption patterns, and ambient air quality are among the factors significantly affected by UHI intensification. As a result of the height and arrangement of urban buildings, urban areas often experience heat waves and intense winds at a height below the average building height. Materials used in pavement in downtown areas have a significant impact on differences in daytime temperature in urban environments. According to research, albedo of urban pavements reduces the intensity of urban heat island during the day, but it is not significant during the night [29]. As well as natural heating systems in urban areas, human-caused heat emissions such as those from transportation and added heating from home activities like cooking and air conditioning contribute significantly to local temperature [30]. In urban areas, the rapid increase in primary air pollution levels is due to the steadily increasing population of vehicles. According to the study, the most important pollutants released from vehicular exhausts are SOx, NOx, CO, PM, and Lead

(Pb) [31]. Mercury, cadmium, nickel, and other heavy metal pollutants also detected in these emissions. In major cities, the population has increased by over 10 million due to rapid industrialization and technological advances [32]. Increasing population density and heavy traffic demands are transforming these regional megacities into complex pollution hotspots [9].

To control air pollution levels in cities, the government sets standards and implements the Urban Air Quality Management Plan (UAQMP). In 1963, the Clean Air Act initiated the first air quality management plan in the USA. Many communities have been identified as non-attainment areas, which continue to suffer from unhealthy pollution above critical levels, by the United States Environmental Protection Agency (USEPA) [33]. India established its air quality monitoring network in 1981 after passing the Air Act. A set of guidelines issued by the Indian Central Pollution Control Board (CPCB) for the preparation of action plans for sixteen cities. An urban environment polluted and contained high concentrations of greenhouse gases. Since the mid-2000s, SOx emissions from traffic are declining while NOx emissions from the same sources are increasing [34]. Recent studies report an increase of 1.5 times in lung cancer risk in an urban area [35].

There are factors responsible for secondary pollution formation, including atmospheric photochemical reactions, primary emissions, meteorology, and solar radiation. Secondary pollution formations in a street canyon should be addressed regionally rather than on an urban scale [36]. Based on the land use pattern, it is important to know the pollutant transformation at a regional level. By releasing reactive pollutants into the air and reacting with the atmosphere, they generate O_3 and other secondary pollutants. The dispersion and transformation of pollutants extensively studied using numerical models with reactive dispersion modeling. A number of air quality models have been used to predict the expected changes in air quality and pollutant levels on a regional scale. Building geometry, land use patterns, and vehicular movement are all necessary parameters for air quality modeling. The formation of secondary particulates often influenced by the temperature.

6. UHI and Thermal Discomfort Tools to Model

The intensity of UHI generated and the level of outdoor thermal comfort significantly influenced by localized surface modification in urban areas. The ambient air temperature, humidity, wind velocity, mean radiant temperature, and other factors affect the outdoor thermal comfort. The heat level in urban hotspots is elevated by factors such as land degradation, impermeable surfaces, and heat emissions [37, 38]. In hotspots, UHI intensity and outdoor thermal discomfort significantly increase cooling energy consumption. There was a report that the energy consumed by urban buildings in London was 25% greater than that of rural buildings [39]. The intensity of thermal discomfort leads to heat-stress related admissions in hospitals [40]. Temperature-related illnesses associated with climate change and outdoor thermal comfort require assessment. In an urban microenvironment, changing the local meteorology and boundary layer structure with increased UHI strengthens ozone formations and affects air quality [41]. The effect of microclimate on outdoor thermal comfort first reported in 1971. Few studies have discussed the urban characteristics and infrastructure effects on thermal comfort levels [21].

The optimization of land cover change in city planning has made extensive use of microclimatic models in recent years. To understand the spatial and temporal variations of outdoor climatological conditions, town energy balance (TEB) models have been used. In the past several years, transient models have become more important for estimating thermal comfort. Models for estimating local meteorological parameters and physiological parameters at variable vegetation levels combined with Weather Research and Forecasting (WRF) and urban climate models (UCM). Designing for outdoor thermal comfort in urban areas is an important factor in how to achieve better thermal comfort [21, 42]. The local thermal sensation simulated using microclimatic models such as radiation models, Rayman, solar and longwave environmental irradiance geometry (SOLWEIG), and CFD and thermodynamic models. In urban zones, the ratio of height to width of buildings also significantly moderated by the upstream and downstream conditions of wind directions. Determining the outdoor thermal comfort and UHI intensity based on the land use pattern and meteorology is a critical part of any research in this area.

In addition to affecting local air quality, hotspots will also cause other heat-related health issues. In addition to urban hotspots, they can also influence climate at the regional level. At the present time, city planning does not consider the intensity variations of urban heat islands (UHIs) or outdoor thermal comfort parameters. Due to rising temperatures and humidity levels, outdoor comfort will become unsustainable in a tropical climate in the future.

7. Urban Characteristics and UHI

7.1. Urban Microenvironment Structure

There are empirical relationships for determining the intensity of urban heat island effects, based on cloud cover, wind speed, population, etc. the nocturnal urban-rural temperature difference, D of the form D = (a-b.N)/v where N is the percent cloud cover, v is the wind speed and a and b are empirically determined constants. Ludwig and Kealoha, (1968) made extensive investigations of the urban heat island effect and developed equations for the determination of the magnitude of urban heat island as a function of variation of temperature with height in the nearby urban areas [43]. An equation for obtaining the maximum value of urban heat island as a function of the population [10]. The heat island effect in relation to the pollution distribution in the city of Toronto [44]. An analysis of the urban heat island effect in New York City conducted by measuring the temperature differentials through the lowest 700m of the atmosphere before and after sunrise in and around the city. In comparison with the surrounding non-urban areas, urban temperature inversions are less intense and less frequent. An urban heat island's average intensity is highest near the surface and decreases to zero at three hundred meters. Heat islands extend well over 500m in the morning with relatively strong urban elevation inversions [45]. The sensitivity of the urban heat island circulation in St. Louis, Missouri, was determined by performing a sensitivity analysis. An equation model of three-dimensional heat island circulation used to study the effects of surface roughness, horizontal diffusion, and boundary layer stability on the heat island circulation. Heat island circulation dominated by the heat island intensity and the boundary layer stability. The number of people migrating from rural and semi-urban areas is causing rapid growth in most Indigenous cities. 80 - 85% of the developed cities in India have asphalt or concrete pavement, with about 30% asphalt and 50% concrete. Concrete and asphalt are low-albedo materials, which reflect solar radiation efficiently. Furthermore, in cities without vegetation, the UHI intensity increases as shadows not cast, evapotranspiration not achieved, and sunlight not reflect as well (due to albedo). Solar radiations received by surface materials are less than those emitted by the surface. The energy balancing equation for each surface material is given by [46]–[48]:

$$Q_{sw,net} + Q_{lw,net}(To) - G(T_o) - H(T_o) - LE(T_o, q_o) = 0 \quad (2.1)$$

Where $Q_{sw, net}$ = net shortwave radiation at the surface, $Q_{lw,net(To)}$ = net long wave radiation at the surface, G = soil heat flux, H = sensible heat flux and LE = latent heat flux. Many studies conducted for determining UHI intensity have used microclimatic models in the past. To balance the energy use of different land uses, the models used only the thermophysical properties of the different urban surface materials.

Micro-scale modeling widely used to determine the intensity of UHI using energy balancing models. UHI is not commonly determined by coupled energy balancing and velocity component models [49]. Temperature and pollution levels significantly affected by seasonal variations in meteorological parameters. There is a substantial difference in mixing height depending on how fast the wind blows, the pressure gradient, and other local conditions. Microscale weather characteristics like temperature, humidity, and mixing height will be affected by land use change [50]. To capture the UHI intensity with land use change in three locations in Chennai city, four velocity conditions for each location simulated.

7.2. UHI Formation Scenarios

There is a difference in roughness and thermal characteristics between urban areas and rural ones. Temperature profiles also affected by severity, but the extent of the effects is determined by thermal features. In cities, asphalt and concrete dominate the landscape. Plants and soil are naturally heat-storage materials. These materials have a higher heat-storing capacity than natural land covers. This causes urban areas to absorb the sun's radiation throughout the day and become warmer than rural areas. Urban surroundings with stable atmospheric conditions remain significantly warmer at night due to the reradiation of stored heat. The rapid industrialization of the twenty-first century and the availability of better infrastructure has resulted in more than half of the world's population living in urban areas. According to estimates, by 2030 over 80% of the population will reside in urban areas. Scientists believe that scientific inventions and discoveries have contributed to the growth of urban centers. In most developing countries, the urban sprawl has already extended beyond the conventional limits of cities; the deliberate and unplanned development of the city into the countryside is referred to as urban sprawl [51].

Buildings and transportation services are the major components of the urban system. The climatic conditions in a metropolitan region affected by the local materials available. The majority of land covers in typical urban areas have low albedo materials, which are characterized by excellent heat transfer from dry surfaces [52]. As a result, land use patterns change, which forces the atmospheric moisture content to decrease. With respect to local temperature rise, low albedo has a significant impact on urban heat island effects. Rural albedo is higher than urban. A typical urban albedo measures 0.25, while a rural albedo measures 0.15. This shows that the net heat absorption in a metropolitan area is higher than in rural areas [53]. Temperature profiles on a local scale will affect by changes in land cover in the urban area. Most of the studies in literature have found that there is a significant difference in temperature between urban and rural areas. Due to an increase in energy demand, pollution, and surface modifications, the downtown area experiences higher temperatures, according to the USEPA (U.S. Environmental Protection Agency). Particularly at night, urban areas have warmer air temperatures than rural areas. The intensity of UHI referred to as the urban heat island intensity (UHII). During the daytime, UHI caused by the surface heat island effect, ranging from 10 to 15°C, whereas at night, the magnitude caused by the atmospheric heat island effect, ranging from 7 to 12°C. UHI contributes to climatic conditions locally because of its intensity.

7.3. Effect of Materials on UHI

Urban heat island development also affected by the thermophysical properties of urban materials, such as solar reflectance, thermal emissivity, and heat capacity. A sunspot's properties determine how its energy is reflected, emitted, or absorbed. Approximately 5% of solar energy is ultraviolet rays, 43% is visible light, and the remaining 52 percent is infrared, which feels as heat. Each of these components reaches the Earth in different proportions: ultraviolet rays comprise 5% of solar energy, visible light 43%, and infrared heat 52%. Each wavelength contributes to the creation of urban heat islands [52]. The albedo or solar reflectance of a surface indicates how much solar energy reflected. Visible wavelengths are responsible for most of the sun's energy. A material's color and solar reflectance are therefore related. The reflectance of dark surfaces is lower than that of light surfaces. Scientists are studying and developing cooling-colored elements that use pigments that are engineered to reflect well infrared rays [54]. Solar reflectance of these products is equivalent to that of white or light-colored materials, even when they are dark in color. Buildings and paving in urban areas typically have a lower albedo than those in rural areas. Because of this, densely populated urban communities often reflect less heat energy and absorb more. This absorbed heat increases surface temperatures, creating urban heat islands on the ground and in the atmosphere. Solar reflectance is the primary determinant of a material's temperature, but its emissivity also contributes. The surface temperature of coatings with high emissivity values remains cooler than that of coatings with low emissivity values. In general, construction materials have high thermal emitting values, with metal being an exception. These properties should consider by manufacturers of cool metallic roofs. Another factor affecting the development of heat islands is the heat capacity of materials. Steel and stone, two common rural building materials, have higher heat capacities than dry soils and sand. Because of this, cities tend to be better at storing solar energy as heat within their infrastructures. When compared to their rural surroundings, metropolitan areas absorb and save twice as much heat during the day. Buildings have the advantage of warming up rapidly and storing much more heat than natural covers during the day [55]. Albedo of the roofing materials was observed to have significant effect on the noontime temperature rise in urban areas [53]. Vegetation index (NDVI) widely used both in rural and urban areas to measure differences in surface properties such as specific heat storage capacity and evaporation. Town Energy Balance (TEB) models used to study the impact of land cover change on urban surface energy balance. Vegetation and urban surfaces exchange energy through different mechanisms. Urban boundary layer thickness depends on the surface roughness factor and built-up area in the urban region. Urban areas have a 300-meter higher atmospheric boundary layer than rural ones.

7.4. Classification of UHI's

In general, UHI can classified into two categories: Surface Heat Island (SHI) and Atmospheric Heat Island (AHI). On a sweltering day in the summer, urban surfaces exposed to the heat of the sun, such as roofs and pavements, have temperatures as high as 40 - 50°C, which is hotter than the surrounding atmosphere. Meanwhile, most rural surroundings remain warm even when shaded or moist. It is common to find urban heat islands on the surface during both day and night, but they tend to be strongest during the day when the sun is shining. In addition to weather and ground cover, the intensity of the sun affects the size of urban heat islands. Because of this variation, summer is typically the best time to observe urban heat islands.

As urban infrastructure slowly releases heat after sunset, urban heat islands can be weak in the morning and throughout the day. They become more prominent after sunset as the heat from urban infrastructure slowly escapes. In addition to these factors, the timing of this peak varies with the properties of the surface, the season, and the prevailing weather.

Atmospheric urban heat islands defined as areas of warmer air in urban areas compared to cooler air in nearby rural surroundings. There are two types of heat islands known as canopy layer and boundary layer [56]. There is air between roughness components in the canopy layer and boundary layer heat island is situated above the canopy layer. Depending on where you live in California (USA), every 10 C drop in summer temperatures decreases electricity consumption by 2-3%, which can reduce carbon dioxide emissions by 10 million tons. There are two distinct layers in the urban atmosphere, the urban canopy layer (UCL) and the urban boundary layer (UBL). There is an area called the UCL between the surface of the earth and the height of buildings; above it is the UBL, the general atmosphere. With temperature measurements made at the standard meteorological height, UCL can measure. Temperature sensors mounted on tall towers or tethered balloons can use to observe UBL heat islands.

7.5. UHI Measuring Techniques

Several methods are used to identify urban heat islands, including direct and indirect methods, numerical modeling, and empirical estimates. Researchers often estimate surface temperatures using remote sensing, a technique of indirect measurement. For evaluating UHI intensity, they create thermal images from the data collected at study locations. First documented in 1818, UHI is a direct consequence of urbanization. A changing land use pattern and air quality (OAQ) affect the city's health. An individual city's UHI intensity is dependent on its location and characteristics. Depending on its location, UHI can take place at the surface or at the atmosphere. Different methods can used to measure each of these, including multiscale phenomena, observable approaches, and simulation methods.

Several approaches can be used to measure surface UHI, including surface temperature monitoring of different landcovers in an urban area, remote sensing, and land surface monitoring. It is possible to measure the UHI using a ground-based temperature-monitoring network or a vehicle-mounted navigator with a thermometer. Remote sensing techniques are another way to estimate UHI. This can be done in a few ways, including:

An urban thermal pattern measure spatially and compared to the characteristics of the surface using thermal remote sensing. A second application was to analyze the urban surface energy balances using thermal remote sensing. As a third technique, thermal remote sensing used to study the relationship between atmospheric heat islands and surface urban heat islands. urban heat island. For ground-level measurements combined with remote sensing, empirical models were developed. On the other hand, they had better results when they measured the UHI using the vegetation index (NDVI) rather than population density.

The decrease in Diurnal Temperature Range (DTR) is an indication of local climate change due to urbanization. As the DTR decreases, the nighttime temperature in the urban region increases.

$$\text{DTR} = (T_{max} - T_{min}) \tag{2.2}$$

Where T_{max} is the maximum diurnal temperature in the daytime.

T_{min} is the minimum diurnal temperature in the nighttime.

Surface temperature of material is a function of albedo and emissivity, the temperature can estimate using the following equation [52]

$$(1 - a)I = \sigma\varepsilon\,(T_{4s} - T_{4sky}) + hc\,(T_s - T_a)$$

Model for UHI calculation	Equation	Reference
Surface energy balance	$Q^{*} + Q_F = Q_H + Q_E + \Delta Q_S$	[57]
Radiation transfer and surface- heat budget	$R_n = V + IE + A$ $C\gamma \frac{\partial T}{\partial x} = \lambda \frac{\partial^2 T}{\partial x^2}$	[58]

Here Q_H = turbulent heat flux, Q_E = latent heat flux, ΔQ_S = storage heat flux, Q^{*} = net all wave radiation, Q_F = anthropogenic heat flux.

R_n = net radiation, A= conduction heat flux,, V= sensible heat flux, IE= latent heat flux, C = heat capacity, λ = is estimated from the observed heat flux, T= temperature.

Conclusion

Study of the urban heat island effect and subsequent meteorological effects on ambient air temperature based on a combination of techniques, ranging from modelling to direct measurements. Various land use patterns have been studied for the intensity and characteristics of UHI. This study indicates that wind velocity and local infrastructure are crucial factors affecting UHI intensity. Land cover and land use patterns affected UHI intensity at specific locations. The intensity of UHI and the formation of heat waves interact complexly in tropical climates. This interaction influenced by a number of factors. Characterizing UHI prediction models requires consideration of these parameters. Its prediction accuracy increased by including both sea breeze and land breeze mechanisms.

Researchers have explored how infrastructure can affect local meteorology and local air quality. According to the monitoring of local temperature and air quality, air quality varies significantly among study locations in the city during different seasons. Local meteorology significantly get influenced by sea breeze and land breeze at this locator future urban expansion, the height of buildings and the wind velocity conditions are crucial.

References

[1] Torres, B., O. J. Maza, P. Aguirre, L. Hinojosa, and S. Günter, "The contribution of traditional agroforestry to climate change adaptation in the Ecuadorian Amazon: The chakra system," in *Handbook of Climate Change Adaptation*, Springer, Berlin, Heidelberg, 2015, pp. 1973–1994. doi: 10.1007/978-3-642-38670-1_90.

[2] Liu, W., J. Zhan, F. Zhao, H. Yan, F. Zhang, and X. Wei, "Impacts of urbanization-induced land-use changes on ecosystem services: A case study of the Pearl River Delta Metropolitan Region, China," *Ecol. Indic.*, 2019, doi: 10.1016/j.ecolind.2018.10.054.

[3] Swamy, G. S. N. V. K. S. N., S. M. Nagendra, and U. Schlink, "Impact of urban heat island on meteorology and air quality at microenvironments," *J. Air Waste Manag. Assoc.*, pp. 876–891, Sep. 2020, doi: 10.1080/10962247.2020.1783390.

[4] Cofala, J., M. Amann, Z. Klimont, K. Kupiainen, and L. Höglund-Isaksson, "Scenarios of global anthropogenic emissions of air pollutants and methane until 2030," *Atmos. Environ.*, vol. 41, no. 38, pp. 8486–8499, Dec. 2007, doi: 10.1016/j.atmosenv.2007.07.010.

[5] Feng, J. L., Z. G. Guo, T. R. Zhang, X. H. Yao, C. K. Chan, and M. Fang, "Source and formation of secondary particulate matter in PM 2.5 in Asian continental outflow," *J. Geophys. Res. Atmos.*, vol. 117, no. D3, p. n/a-n/a, Feb. 2012, doi: 10.1029/2011JD016400.

[6] Ning, Z., C. Cheung, Y. Lu, M. Liu, and W. Hung, "Experimental and numerical study of the dispersion of motor vehicle pollutants under idle condition," *Atmos. Environ.*, vol. 39, no. 40, pp. 7880–7893, Dec. 2005, doi: 10.1016/j.atmosenv.2005.09.020.

[7] Willers, S. M., M. F. Jonker, L. Klok, M. P. Keuken, J. Odink, S. van den Elshout, C. E. Sabel, J. P. Mackenbach and A. Burdorf. "High resolution exposure modelling of heat and air pollution and the impact on mortality," *Environ. Int.*, vol. 89–90, pp. 102–109, Apr. 2016, doi: 10.1016/j.envint.2016.01.013.

[8] Guo, X., Y. Sun, and S. Miao, "Characterizing Urban Turbulence Under Haze Pollution: Insights into Temperature–Humidity Dissimilarity," *Boundary-Layer Meteorol.*, vol. 158, no. 3, pp. 501–510, Mar. 2016, doi: 10.1007/s10546-015-0104-y.

[9] Baklanov, A., L. T. Molina, and M. Gauss, "Megacities, air quality and climate," *Atmos. Environ.*, vol. 126, no. 3, pp. 235–249, Feb. 2016, doi: 10.1016/j.atmosenv. 2015.11.059.

[10] Tayanc, M. and H. Toros, "Urbanization effects on regional climate change in the case of four large cities of turkey," *Clim. Change*, vol. 35, pp. 501–524, 1997, doi: https://dx.doi.org/10.1023/A:1005357915441.

[11] Fenger, J. "Urban air quality," *Atmos. Environ.*, vol. 33, no. 29, pp. 4877–4900, Dec. 1999, doi: 10.1016/S1352-2310(99)00290-3.

[12] Gulia, S., S. M. Shiva Nagendra, M. Khare, and I. Khanna, "Urban air quality management-A review," *Atmos. Pollut. Res.*, vol. 6, no. 2, pp. 286–304, Mar. 2015, doi: 10.5094/APR.2015.033.

[13] Chemke, R., Y. Kaspi, and I. Halevy, "The thermodynamic effect of atmospheric mass on early Earth's temperature," *Geophys. Res. Lett.*, vol. 43, no. 21, pp. 11,414-11,422, Nov. 2016, doi: 10.1002/2016GL071279.

[14] Solecki, W. and C. Rosenzweig, "Indicators and monitoring systems for urban climate resiliency," *Clim. Change*, vol. 163, no. 4, pp. 1815–1837, Dec. 2020, doi: 10.1007/s10584-020-02947-4.

[15] Fong, C. S., N. Aghamohammadi, L. Ramakreshnan, N. M. Sulaiman, and P. Mohammadi, "Holistic recommendations for future outdoor thermal comfort assessment in tropical Southeast Asia: A critical appraisal," *Sustain. Cities Soc.*, vol. 46, no. January, p. 101428, Apr. 2019, doi: 10.1016/j.scs.2019.101428.

[16] Gonzalez-Trevizo, M. E., K. E. Martinez-Torres, J. F. Armendariz-Lopez, M. Santamouris, G. Bojorquez-Morales, and A. Luna-Leon, "Research trends on environmental, energy and vulnerability impacts of Urban Heat Islands: An overview," *Energy and Buildings*, vol. 246. Elsevier Ltd, Sep. 01, 2021. doi: 10.1016/j.enbuild.2021.111051.

[17] Wheeler, S. "A New Conception of Planning in the Era of Climate Change," *Berkeley Plan. J.*, vol. 23, no. 1, Nov. 2011, doi: 10.5070/BP32318536.

[18] Caiado, R. G. Gusmão, W. Leal Filho, O. L. G. Quelhas, D. Luiz de Mattos Nascimento, and L. V. Ávila, "A literature-based review on potentials and constraints in the implementation of the sustainable development goals," *J. Clean. Prod.*, vol. 198, pp. 1276–1288, Oct. 2018, doi: 10.1016/J.JCLEPRO.2018.07.102.

[19] Rizwan, A. M., L. Y. C. Dennis, and C. Liu, "A review on the generation, determination and mitigation of Urban Heat Island," *J. Environ. Sci.*, vol. 20, no. 1, pp. 120–128, Jan. 2008, doi: 10.1016/S1001-0742(08)60019-4.

[20] Memon, R. A., D. Y. C. Leung, and C.-H. Liu, "An investigation of urban heat island intensity (UHII) as an indicator of urban heating," *Atmos. Res.*, vol. 94, no. 3, pp. 491–500, Nov. 2009, doi: 10.1016/j.atmosres.2009.07.006.

[21] Salata, F., I. Golasi, R. de Lieto Vollaro, and A. de Lieto Vollaro, "Urban microclimate and outdoor thermal comfort. A proper procedure to fit ENVI-met simulation outputs to experimental data," *Sustain. Cities Soc.*, vol. 26, pp. 318–343, Oct. 2016, doi: 10.1016/j.scs.2016.07.005.

[22] Hare, F. K. and H. E. Landsberg, "The Urban Climate," *Geogr. Rev.*, vol. 73, no. 2, p. 242, Apr. 1983, doi: 10.2307/214663.

[23] Voogt, J. a. and T. R. Oke, "Thermal remote sensing of urban climates," *Remote Sens. Environ.*, vol. 86, no. 3, pp. 370–384, 2003, doi: 10.1016/S0034-4257(03) 00079-8.

[24] Gala, T. S., R. Alfraihat, G. Mulugeta, and T. S. Gala, "Ecological Evaluation of Urban Heat Island in Chicago City, USA Environmental Monitoring View project GIS and Public Health View project Ecological Evaluation of Urban Heat Island in Chicago City, USA," *J. Atmos. Pollut.*, vol. 4, no. 1, pp. 23–29, 2016, doi: 10.12691/jap-4-1-3.

[25] Jones, P. D. and P. a. Reid, "Temperature trends in regions affected by increasing aridity/humidity," *Geophys. Res. Lett.*, vol. 28, no. 20, pp. 3919–3922, Oct. 2001, doi: 10.1029/2001GL013840.

[26] Fujibe, F. "Urban warming in Japanese cities and its relation to climate change monitoring," *Int. J. Climatol.*, vol. 31, no. 2, pp. 162–173, Feb. 2011, doi: 10.1002/joc.2142.

[27] Goodridge, J. D. "Urban bias influences on long-term California air temperature trends," *Atmos. Environ. Part B. Urban Atmos.*, vol. 26, no. 1, pp. 1–7, Mar. 1992, doi: 10.1016/0957-1272(92)90032-N.

[28] Memon, R. A. and D. Y. C. Leung, "Impacts of environmental factors on urban heating," *J. Environ. Sci.*, vol. 22, no. 12, pp. 1903–1909, Dec. 2010, doi: 10.1016/S1001-0742(09)60337-5.

[29] Radhi, H., S. Sharples, and E. Assem, "Impact of urban heat islands on the thermal comfort and cooling energy demand of artificial islands—A case study of AMWAJ Islands in Bahrain," *Sustain. Cities Soc.*, vol. 19, pp. 310–318, Dec. 2015, doi: 10.1016/j.scs.2015.07.017.

[30] Eliasson, I. "Urban nocturnal temperatures, street geometry and land use," *Atmos. Environ.*, vol. 30, no. 3, pp. 379–392, 1996, doi: 10.1016/1352-2310(95)00033-X.

[31] Amoatey, P. H. Omidvarborna, M. S. Baawain, and A. Al-Mamun, "Emissions and exposure assessments of SOX, NOX, PM10/2.5 and trace metals from oil industries: A review study (2000–2018)," *Process Safety and Environmental Protection*, vol. 123. Institution of Chemical Engineers, pp. 215–228, Mar. 01, 2019. doi: 10.1016/j.psep.2019.01.014.

[32] Hu, H., Q. Jin, and P. Kavan, "A Study of Heavy Metal Pollution in China: Current Status, Pollution-Control Policies and Countermeasures," *Sustain. 2014, Vol. 6, Pages 5820-5838*, vol. 6, no. 9, pp. 5820–5838, Sep. 2014, doi: 10.3390/ SU6095820.

[33] Orford, A. D. "The Clean Air Act of 1963: Postwar Environmental Politics and the Debate over Federal Power," *Hast. Environ. Law J.*, vol. 27, 2021, Accessed: Mar. 15, 2022. [Online]. Available: https://heinonline.org/HOL/Page?handle=hein.journals/haswnw27&id=298&div=&collection=

[34] Cullinane, K. and S. Cullinane, "*Atmospheric Emissions from Shipping: The Need for Regulation and Approaches to Compliance,*" http://dx.doi.org/10.1080/01441647.2013.806604, vol. 33, no. 4, pp. 377–401, Jul. 2013, doi: 10.1080/01441647.2013.806604.

[35] Nyberg, F., P. Gustavsson, L. Jarup, T. Bellander, K. Berglind, R. Jakobsson and G. Pershagen. "Urban Air Pollution and Lung Cancer in Stockholm," *Epidemiology*, vol. 11, no. 5, pp. 487–495, Sep. 2000, doi: 10.1097/00001648-200009000-00002.

[36] Liu, C. H. and D. Y. C. Leung, "Numerical study on the ozone formation inside street canyons using a chemistry box model," *J. Environ. Sci.*, vol. 20, no. 7, pp. 832–837, 2008, doi: 10.1016/S1001-0742(08)62134-8.

[37] Santamouris, M. "Cooling the cities – A review of reflective and green roof mitigation technologies to fight heat island and improve comfort in urban environments," *Sol. Energy*, vol. 103, pp. 682–703, May 2014, doi: 10.1016/j.solener.2012.07.003.

[38] Baik, J.-J., Y.-S. Kang, and J.-J. Kim, "Modeling reactive pollutant dispersion in an urban street canyon," *Atmos. Environ.*, vol. 41, no. 5, pp. 934–949, Feb. 2007, doi: 10.1016/j.atmosenv.2006.09.018.

[39] Hart, M. a. and D. J. Sailor, "Quantifying the influence of land-use and surface characteristics on spatial variability in the urban heat island," *Theor. Appl. Climatol.*, vol. 95, no. 3–4, pp. 397–406, Mar. 2009, doi: 10.1007/s00704-008-0017-5.

[40] Taleghani, M., M. Tenpierik, S. Kurvers, and A. van den Dobbelsteen, "A review into thermal comfort in buildings," *Renew. Sustain. Energy Rev.*, vol. 26, pp. 201–215, Oct. 2013, doi: 10.1016/j.rser.2013.05.050.

[41] Bell, M. C., F. Galatioto, A. Chakravartty, and A. Namdeo, "A novel approach for investigating the trends in nitrogen dioxide levels in UK cities," *Environ. Pollut.*, vol. 183, no. 2, pp. 184–194, Dec. 2013, doi: 10.1016/j.envpol.2013.03.039.

[42] Mohan, M. and A. Kandya, "Impact of urbanization and land-use/land-cover change on diurnal temperature range: A case study of tropical urban airshed of India using remote sensing data," *Sci. Total Environ.*, vol. 506–507, pp. 453–465, Feb. 2015, doi: 10.1016/j.scitotenv.2014.11.006.

[43] Shi, Y, Y. Xiao, Z. Li, X. Zhang, T. Liu, Y. Li, Y. Pan and W. Yan. "Microorganism structure variation in urban soil microenvironment upon ZnO nanoparticles contamination," *Chemosphere*, vol. 273, p. 128565, Jun. 2021, doi: 10.1016/J.CHEMOSPHERE.2020.128565.

[44] Jauregui, E. "Heat island development in Mexico City," *Atmos. Environ.*, vol. 31, no. 22, pp. 3821–3831, Nov. 1997, doi: 10.1016/S1352-2310(97)00136-2.

[45] Arnfield, A. J. "Two decades of urban climate research: a review of turbulence, exchanges of energy and water, and the urban heat island," *Int. J. Climatol.*, vol. 23, no. 1, pp. 1–26, Jan. 2003, doi: 10.1002/joc.859.

[46] Du, H., D. Wang, Y. Wang, X. Zhao, F. Qin, Hong Jiang and Y. Cai. "Influences of land cover types, meteorological conditions, anthropogenic heat and urban area on surface urban heat island in the Yangtze River Delta Urban Agglomeration," *Sci. Total Environ.*, vol. 571, pp. 461–470, 2016, doi: 10.1016/j.scitotenv.2016.07.012.

[47] Yang, X., Y. Chen, L. L. H. Peng, and Q. Wang, "Quantitative methods for identifying meteorological conditions conducive to the development of urban heat islands," *Build. Environ.*, vol. 178, no. May, p. 106953, 2020, doi: 10.1016/j.buildenv.2020.106953.

[48] Wang, T., L. Xue, P. Brimblecombe, Y. F. Lam, L. Li, and L. Zhang, "Ozone pollution in China: A review of concentrations, meteorological influences, chemical precursors, and effects," *Sci. Total Environ.*, vol. 575, pp. 1582–1596, Jan. 2017, doi: 10.1016/j.scitotenv.2016.10.081.

[49] Toparlar, Y., B. Blocken, P. Vos, G. J. F. van Heijst, W. D. Janssen, T. van Hoof, H. Montazeri, and H. J. P. Timmermans. "CFD simulation and validation of urban microclimate: A case study for Bergpolder Zuid, Rotterdam," *Build. Environ.*, vol. 83, pp. 79–90, Jan. 2015, doi: 10.1016/J.BUILDENV.2014.08.004.

[50] Yang, X., L. Zhao, M. Bruse, and Q. Meng, "Evaluation of a microclimate model for predicting the thermal behavior of different ground surfaces," *Build. Environ.*, vol. 60, pp. 93–104, Feb. 2013, doi: 10.1016/j.buildenv.2012.11.008.

[51] Tayanç, M. and H. Toros, "Urbanization effects on regional climate change in the case of four large cities of Turkey," *Clim. Change*, vol. 35, no. 4, pp. 501–524, 1997, doi: 10.1023/A:1005357915441.

[52] Prado, R. T. A. and F. L. Ferreira, "Measurement of albedo and analysis of its influence the surface temperature of building roof materials," *Energy Build.*, vol. 37, no. 4, pp. 295–300, 2005, doi: 10.1016/j.enbuild.2004.03.009.

[53] Taha, H., D. Sailor, and H. Akbari, "High-Albedo Materials for Reducing Building Cooling Energy Use.," 1992.

[54] Salata, F., I. Golasi, A. D. L. Vollaro, and R. D. L. Vollaro, "How high albedo and traditional buildings' materials and vegetation affect the quality of urban microclimate. A case study," *Energy Build.*, vol. 99, pp. 32–49, 2015, doi: 10.1016/j.enbuild.2015.04.010.

[55] Gray, K. A. "The Urban Heat island, Photochemical Smog," *Depatment Civ. Eng. Northwest. Univ. Evast.*, 2000.

[56] Voogt, J. a. and C. S. B. Grimmond, "Modeling Surface Sensible Heat Flux Using Surface Radiative Temperatures in a Simple Urban Area," *J. Appl. Meteorol.*, vol. 39, pp. 1679–1699, 2010, doi: 10.1175/1520-0450(2000)039<1679:MSSHFU>2.0.CO;2.

[57] Grimmond C. S. B. and M. Blackett, "The International Urban Energy Balance Models Comparison Project : First Results from Phase 1," pp. 1268–1292, 2010, doi: 10.1175/2010JAMC2354.1.

[58] Takebayashi, H., Y. Kimura, and S. Kyogoku, "Study on the appropriate selection of urban heat island measure technologies to urban block properties," *Sustain. Cities Soc.*, vol. 13, pp. 217–222, Oct. 2014, doi: 10.1016/j.scs.2014.01.008.

Chapter 2

Urban Heat Island and Mitigation Strategies: International Studies and Current Status in India

Jagabandhu Panda[1,*], Asmita Mukherjee[1] and Sudhansu S. Rath[2]
[1]Department of Earth and Atmospheric Sciences,
National Institute of Technology Rourkela, Odisha, India
[2]CSIR Fourth Paradigm Institute, Bangalore, India

Abstract

Urbanization results in rising temperature over urban areas forming an 'urban heat island (UHI)' phenomenon. Cities worldwide are facing the growing problem of the UHI effect. Internationally, the UHI intensity of up to 12°C and up to 9°C in Indian cities is observed. Although the numerical modeling studies are limited in this sense, simulations suggest up to 5°C UHI intensity over medium to large-size Indian cities. The sustainable development goals at an urban scale can be achieved by improving the UHI mitigation measures. Although UHI and its impacts have been studied extensively over major cities globally, it is relatively less over India. The UHI mitigation technologies are under development, and application is minimal. The various mitigation measures, technologies, and their efficiencies need to be transferred to the designers and planners for practicing sustainable city developments. The current review intends to briefly present the UHI scenarios and available technologies for mitigation measures worldwide. A systematic review of actual developments of UHI mitigation over Indian cities is presented

[*]Corresponding Author's Email: jagabandhu@gmail.com or pandaj@nitrkl.ac.in.

In: Urban Heat Islands Reexamined
Editors: Satyaprakash and Anne W.M. Ng
ISBN: 979-8-88697-215-3

with the possible suggested adaptation process. Adopting suitable UHI mitigation measures requires a specific background understanding of the local climate zone (LCZ) besides the inherent UHI phenomena, which is found limited over Indian cities. Although the knowledge of UHI is taking a leap using the LCZ based studies over India, the mitigation part is yet to be discovered. This review work would possibly help advance the knowledge in the Indian perspectives by aggregating the learnings from international practices.

Keywords: urbanization, UHI, mitigation, LCZ, sustainability

1. Introduction

Climate change issues are considered one of the most pressing problems faced by the world today, among other environmental issues such as global warming, pollution, water, land contamination, etc. Studies related to climate change have shown that Land Surface Temperature (LST) and its metrics have a significant amount of impact and are thus considered as an important parameter for urban heat island (UHI), surface energy budget, and other climate models (Brohan et al. 2006; Xiao et al. 2007; Hansen et al. 2010; Li et al. 2013). Among other causes, urban expansion is one of the major reasons behind the significant change in land use and surface temperature (Argüeso et al. 2015; Hegazy and Kaloop 2015). Urbanization is quite prominent in developed and advanced (technologically and economically) countries. The land use and land cover (LULC) change resulting from urbanization and associated anthropogenic activities contribute significantly to the local and regional weather and climate. Therefore, urbanization has been linked to climate change during the past two decades (Chapman et al. 2017).

Migration from rural to urban areas and the transformation of rural settlements into cities are important determinants of the anticipated high urban population. The world's urban population continues to grow faster than the total population (UN Report, 2014). In 2008, more than half of the world's population was living in urbanized regions (UNPF, 2007), and the urban population is expected to reach 81% by 2030. Cities in the less developed world are expected to absorb almost all of the future population growth. By 2030, Asia and Africa will each have more urban dwellers than any other major area, with Asia alone accounting for more than 50% of the world's urban population. Due to the acceleration of global urbanization (including India) in both intensity and area, there is a growing interest in understanding its implications. In India, rapid urbanization in larger cities attracted more and more rural to urban migration. According to the Census of India 2011, there has been a growth of the urban population by 31.8%. Consequently, urbanization took place with an annual expansion rate of 2.4% from 1992 to 2012 (Gibson et al. 2015), impacting the land use pattern of the cities with a decrease in vegetation cover and an increase in impervious surfaces (Zhang et al. 2021). All of these contribute to the formation of UHI, which generally refers to the phenomenon where the temperature of the urban area becomes higher than its surrounding, forming an island-like scenario (EPA 2008).

Interactions between the atmosphere and the land surface have considerable influences on weather and climate over a region (Avissar 1995; Barros and Hwu 2002; Liang and Xie 2003) in general. When it comes to urban areas, in particular, the interaction is quite important due to significant changes in LULC over a few decades (Comarazamy et al. 2010; Li et al. 2013; Shepherd and Burian 2003; Shepherd 2005). A tropical urban region experiences extreme weather events, including UHI driven by relatively stronger convection. Therefore, it would not be wrong to say that UHI is one of the consequences of land-atmosphere interaction too.

Although it is good to have well-organized, well-planned, and beautiful cities, there would be some impacts of the growing urbanization in the form of UHI and alteration of local weather and climate. This chapter intends to discuss various aspects of UHI, the consequences, and mitigation strategies.

2. Urban Heat Island, Causes, and Its Effects

UHI occurs on the surface and in the atmosphere. Surface UHIs are typically present in day and night time but tend to be strongest during the day when the sun is shining. In contrast, atmospheric UHIs are often weak during the late morning and throughout the day and are usually more pronounced during night. They can be classified into three categories for different layers of the urban environment (Figure 1), viz. Surface Urban Heat Island (SUHI), Canopy Layer Urban Heat Island (CLUHI), and Boundary Layer Urban Heat Island (BLUHI). The UHI generated by urban surfaces such as buildings, roads, pavements, open space, water bodies, and vegetation is known as SUHI. SUHI not only takes into consideration the differences in surface temperature but also the temporal variability (Voogt and Oke 2003). It can be measured using remote sensing sensors mounted on aircraft or satellites. SUHI is the largest in the daytime during the summer season. CLUHI is generated in the Canopy

Layer (or UCL), or the layer of air closest to the various surfaces in a city following the surface layer extending upwards to about the mean building height of a specific location. CLUHI is measured using thermometers fixed on vehicles or by using sensors at a standard meteorological elevation (Voogt and Oke 2003). It is observed to increase during the late afternoon and early evening, reaching the maxima at night while generating a cool island during the daytime. BLUHI forms in the Urban Boundary Layer (UBL), or the layer above UCL, and can reach up to 1 km or slightly higher during the day and hundreds of meters or less at night (Voogt 2004). BLUHI can be measured remotely from ground-based instruments and using thermometers mounted on tall buildings or towers, aircraft, etc. It has been observed that the yearly mean air temperature of a city with a population of 1 million or more people can be 1–3°C warmer than the surrounding landscape (EPA 2008).

There are five main causes for the occurrence of UHI over urban areas. The primary cause is heat production due to industrial activities, cooling and heating for thermal comfort, and the concentration of people (Sailor 2013; EPA 2008; Filho et al. 2021). The second reason is the changes in the composition of the atmosphere due to an increase in greenhouse gases (viz., CO_2, CO), enhancing the greenhouse effect (Perera and Sanford 2011). Due to the industrial and transport activities, turbidity of the air can increase (Mohajerani et al. 2017), which can be a contributor to the occurrence of UHI too. Also, gradual loss of urban forest cover, thermal properties of the materials over the cities, urban morphology and city size, and anthropogenic heat produced over the cities can contribute significantly to the UHI (Filho et al. 2021; Mohajerani et al. 2017). Besides, the developed cities experience more impermeability due to replacing natural soils with such a category of materials, e.g., asphalt and most building materials (EPA 2008; Mirzaei and Haghighat 2010). Thus, UHIs are influenced by the latitude and elevation, urbanization, vegetation cover, LULC, underlying impervious surfaces, prevailing urban canyons, building densities, distribution, and consumption of heating/cooling energy, ventilation, the intensity of road traffic, and air pollution (Tzavali et al. 2015).

UHIs are one of the major concerns for advanced and larger cities of the world. They can affect communities by increasing summertime peak energy demand, air conditioning costs (Li et al. 2018, 2019), air pollution and greenhouse gas emissions (Singh et al. 2019), heat-related illness and mortality (Gamble et al. 2013), and water quality. Besides, they can impact the local or regional weather and climate too (Bornstein 1968; Niyogi et al.

2009). Therefore, the UHI effects are illustrated separately in the following sub-section.

2.1. Effects of UHI

Since the past two decades, climate change and its effects have been prominently observed over major cities of the world, escalating frequent occurrences of heatwaves and temperature anomalies within the urban center as compared to the surrounding areas (Filho et al. 2018). Heat waves, heat stress, and rising temperatures contribute to climate change on a global scale, have adverse effects on human health, and degrade air quality and water availability (Zhao et al. 2018; Venter et al. 2020). The 2019 Lancet countdown report showed a record increase in heatwave instances in 2018, impacting around 220 million people worldwide. Watts et al. (2019) highlighted that heat stress also leads to reduced ability to work. The UHI effect can be considered quite important in the city environment during such circumstances. Notably, the UHI effect may vary with city size, LULC change, topography, urbanization, industrialization, the season of a year, time of the day, and prevailing meteorology (Tzavali et al. 2015). With adverse impacts in the local microclimate (Filho et al. 2017), UHI intensification may contribute to the following consequences:

(a) **Impact on weather and climate:** Cities throughout the world are projected to expand and regenerate, giving rise to the UHI effect occurring as a result of LULC changes. It is found to have an adverse impact on urban microclimate, altering physical properties like albedo, surface roughness, energy fluxes, atmospheric moisture content, etc. leading to variations in weather and climate systems (Bornstein 1968; Niyogi et al. 2009). While analyzing the impacts of UHI, Oke (1995) pointed out that cities are observed to enhance both the mechanical and thermal turbulence due to an increase in the surface heat flux. He also added that the heat anomaly experienced by a city affects the airflow by causing variations in the local pressure field. Sometimes, weaker, calm winds influence the formation of Heat Island Circulation (HIC), which often interacts with other thermal circulations in the case of coastal cities. Therefore, there is a higher probability that during the prevailing hot weather conditions (primarily during summer), the intensity, horizontal distribution, and vertical extension of UHI can get enhanced in the afternoon hours (Zhang et al. 2011) to create more discomfort in the cities. Landsberg (1956) described the first instance of urban-induced modifications in precipitation where there were indications that the city of Tulsa, Oklahoma, was causing more precipitation. Later, Huff and Changnon (1973) concluded that UHI enhances precipitation rather than initiating it. UHI also impacts rainfall patterns because of increased condensation nuclei and surface convergence. As a result, increased precipitation is observed in the downwind urban region (Landsberg 1956; Changnon 1981; Mather 1991). However, more recent studies showed that UHI merely increases the amount of precipitation as urban pollution creates droplets that do not become large enough to fall as precipitation and hence suppress the amount of rainfall in the urban center (Rosenfeld 2000). Bornstein and LeRoy (1990) highlighted the case of New York City (NYC), which affected the formation and movement of thunderstorms during summer daytime. The study also added that thunderstorms approaching the city tend to diverge and move around the rim of the urban area (Niyogi et al. 2011), where UHI was predominant. The works of Dixon and Mote (2003) indicated that the long-prevailing UHIs over a city could initiate thunderstorms and significantly impact the spatial and temporal patterns of rainfall during summer. A well-developed UHI can perturb thermal and dynamical processes over urban areas impacting the location of thunderstorms and precipitation distribution (Lin et al. 2011). From a climatological perspective, the urban microclimate is affected mainly by urban morphology and consistent higher anthropogenic activities. The UHI effect contributes around 30% to climate change (Huang and Lu 2015). On the contrary, an earlier study suggested that UHI had minimal impact on world mean temperature changes (Peterson et al. 1999). Nonetheless, cities can alter the climate of area 2–4 times more than their own (Zhou et al. 2015). And UHI can influence world climate by affecting the jet stream (Zhang et al. 2013). However, climate change can impact the severity of UHI effects, i.e., the intensity of the effect becomes higher in such a scenario (Sachindra et al. 2016; Tuholske et al. 2021).

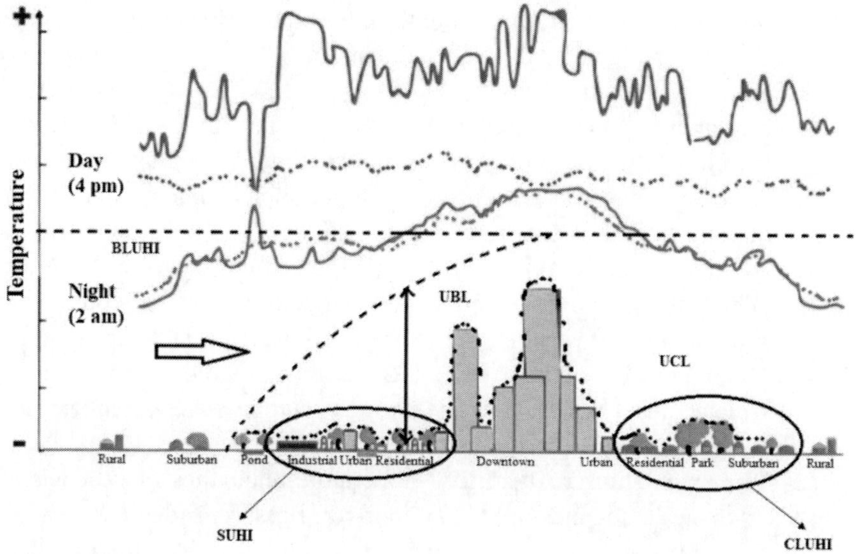

Figure 1. Schematic representation of temperature variation (solid line – surface temperature, dotted line – air temperature) during the day (red color) and night (blue color), different types of UHI, UBL, and UCL. The figure is reproduced using COREL DRAW by following USGS (https://www.usgs.gov/media/images/urban-heat-islands) and Almeida et al. (2021).

(b) **Increasing energy consumption:** UHI is observed to have a major influence on urban building energy usage. Due to the urban heat stress, there is excess demand for electricity, mainly during the summer months, causing power plants to generate extra energy to meet the ever-increasing needs. With rising urban temperature, more and more instances of increase in cooling and decrease in heating demands are found to be observed, leading to frequent outages and blackouts (Maxwell et al. 2018; Lead et al. 2018). UHI might result up to 19% increase or decrease in building cooling or heating energy consumption. However, the lack of high-resolution data makes it difficult to analyze the extent of UHI impact on energy consumption issues (Li et al. 2018, 2019).

(c) **Greenhouse gas emissions and air pollution:** Due to the higher demand for electricity consumption during summer, more and more fossil fuels are consumed, releasing an excessive amount of greenhouse gases and air pollutants into the atmosphere, which further leads to the decline in air quality (Singh et al. 2019). The pollutants that are released into the environment include ground-level ozone or smog and particulate matter. It also leads to the formation of acid rain. Dust and particulate matter are continuously emitted from the anthropogenic activities pertaining to the construction works. Besides, vehicular pollution has burdened the city environment with various primary and secondary air pollutants. Increased UHI can directly increase the rate of ground-level ozone generation, in addition to their influence on energy-related emissions (EPA 2008). When nitrogen oxides and volatile organic molecules combine in sunlight and hot weather, ground-level ozone is generated. Therefore, urban air pollution has been a big concern for the developing world, which must be dealt with through cutting-edge scientific and technological findings.

(d) **Impact on human health:** Temperature anomalies or heat islands often lead to increased day temperatures and decreased night temperatures, with higher rates of air pollution. All these lead to sudden heat-related deaths and illnesses such as discomfort, respiratory difficulties, blood-pressure changes, heat cramps, heat strokes, etc. The older population is more vulnerable to such events, probably due to their immobility, isolated conditions, or lower income levels (Gamble et al. 2013). According to a report by the U.S. Climate Change Science Program (2008), young children are also susceptible to such weather extremities, which might raise their possibility of developing asthmatic conditions or lung diseases at a later stage. Additionally, people with lower income, field-related outdoor jobs, and already diagnosed diseases like diabetes are also considered vulnerable to UHI effects.

(e) **Degrading water quality:** Storm-water runoff heated by hot pavements and rooftop surfaces (occur due to UHI effect) draining into sewers boosts water temperatures as it flows into streams, rivers, ponds, and lakes, impacting millions of aquatic organisms. Warm storm-water runoff can cause rapid temperature swings in aquatic habitats, stressful and even lethal to aquatic species. Somers et al. (2016) mentioned that urban streams are hotter than streams flowing through forested regions with higher water temperatures during minor storms due to heated surface runoff from urban impermeable surfaces.

3. Mitigation Strategies

UHI directly affects human health by creating heat stress, causing acute and chronic exposure to air pollutants, and reducing physical health and well-being. The continuous, direct exposure to extreme UHI can cause adverse health impacts and human discomfort for the city inhabitants. The challenges imposed by the UHI can be dealt with the 'adaptation' and 'mitigation' practices. Adaptation is a short-term adjustment of human behaviors and systems without affecting the natural urban system. The limitation associated with the adaptation measures gives rise to alternatives, termed as mitigation practices. Mitigation included a set of scientifically well-conceived measures or strategies involving the governing bodies and associated stakeholders. According to local circumstances, specific intervention measures vary in scale and magnitude, implementation potential, and suitability.

A lack of a proper classification system for UHI mitigation strategies was noticed while referring to the available literature. Giguère (2009) grouped in detail the mitigation measures into four major categories, which include (i) Greening measures, (ii) Sustainable urban infrastructure, (iii) Sustainable water management, and (iv) Anthropogenic heat reduction. These are prepared according to the application perspective in the real-world condition and are briefly described in the following sub-sections.

3.1. Greening Measures

One of the key measures to counter the UHI phenomenon is by increasing the use of green approaches, including green technologies and vegetation (Arabi et al. 2015). The benefits of green measures are significant over the urban and suburban areas with less or no vegetation. There are several ways to achieve greening measures, i.e., by providing vegetation covers over roofs and walls and by tree plantation in available vacant grounds.

Roofs of private and public buildings can be covered with appropriate vegetation and are proved to be useful in reducing the heat (Block et al. 2012) compared to the conventional roof structures due to the combined effect of evapotranspiration and an additional insulation layer that reduces the heat flux at the roof surfaces. Extensive and intensive green roofing are categorized based on the depth of soil and vegetation layer over the roof. Extensive green roofing is more common because of its easy maintenance, lower cost, and less weight stress on the building structure. Previous research acknowledged green roofs' ability in UHI mitigation (Baniya et al. 2018; Dwivedi and Mohan 2018). Studies like Arabi et al. (2015) recommended using green roofing as a primary strategy for the mitigation of UHI impacts.

Like the green roof, covering the facades or walls of the buildings with vegetation reduces the absorption of shortwave radiation by the building walls (Balany et al. 2020). Green facades differ from green roof technology in terms of their application procedure and type of vegetation used. In this type, climbing plants are set up with the help of a support mechanism like cables, clamps, and metal frames attached to the building. This type of mitigation measure is also known as 'living wall' (Herath et al. 2018)

The primary role of shading trees is to provide shade in the surrounding environment through its spreading canopy and crowns. Shading reduces the incidence of solar radiation on horizontal and vertical surfaces. In addition to shade, trees reduce the near-surface temperature by evaporative cooling and increased relative humidity (Battista et al. 2020). Shading trees mitigation is suitable for parking lots, vegetation around the buildings, urban parks, house yards, road corridors, etc.

Ground vegetation refers to the extensive coverage on the ground level, mainly in the form of grass. Ground vegetation works well in combination with the appropriate amount of irrigation water, which keeps the near-surface air and surrounding environment cooler through evapotranspiration and trapping air near the ground surface, thus improving the microclimate (Balany et al. 2020; Gunawardena et al. 2017). In the absence of proper irrigation, increased humidity can cause human discomfort during the nighttime in a hot and humid climate.

Various studies are performed worldwide considering the above-discussed greening measures to understand the mechanism of UHI and its mitigation. Li et al. (2014) investigated the green roof performance and emphasized the enhanced cooling due to the irrigation-lead soil moisture increment. The city-scale deployment of green vegetation increases the latent heat flux and lowers heat storage during the daytime, and contributes to the lower sensible heat flux, and hence, lower near-surface air temperature. The vegetation measures are preferable for nocturnal UHI mitigation (Li and Norford 2016). The cooling efficiency of parks in mitigating UHI is determined by species group, canopy cover, size, and shape of parks, thus suggesting the planning and designing of urban parks accordingly (Feyisa et al. 2014). Besides the green roof, vertical walls and also dense urban vertical cover or forests are recommended to reduce the effect of UHI as an optimal design (Rosenzweig et al. 2006; Price et al. 2015; Dwivedi and Mohan 2018). The increased greenery works alternatively for summer and winter, thus must be considered as a design element to achieve year-long effectiveness (Chun and Guldmann 2018). The different LULC like vegetation and grass cover of parking lots play different roles while mitigating UHI and have seasonal characteristics. The study revealed tree cover of parking lots has better mitigating capabilities than grass (Takebayashi et al. 2009; Onishi et al. 2010).

Sun and Chen (2017) reported the complex relationship between the number of green spaces and temperature mitigation and suggested focusing on natural forests inside urban areas for effective mitigation. In certain instances, the green roofs mitigate considerably less than the cool roof measures (Yang et al. 2018); thus, mixed measures are suggested for effective mitigation. Studies on the green roof also indicated a marginal effect on UHI mitigation in high-rise urban morphology, while increased grass and tree coverage ratios were found to be more effective and sustainable (Kim et al. 2018). However, the green approaches are not always suitable because of the shortage of available space for greenery establishments. Densely packed urban environments and the high cost of urban lands are obstacles in greening measures.

3.2. Sustainable Urban Infrastructure

Sustainable urban infrastructure measures for UHI mitigation primarily include the modifications of the external envelopes or exposed surfaces using cool building materials. These modifications can reduce the heat absorbed by structures and the heat they release to their surroundings. This can be achieved for a singular building and the whole built environment. UHI on the urban surface depends upon the amount of solar radiation absorbed by the materials used. The building materials having very low albedo absorb considerable solar radiation. Application of materials with very high albedo to the roofs and walls is effective in reducing the temperature, cooling load for buildings, and the amount of heat emitted to the surroundings (Falasca et al. 2019).

Materials used conventionally for pavements like concrete and asphalt absorb a large amount of heat when exposed to direct sunlight throughout the day. These materials emit infrared radiation to the streets and surroundings during the night, which enhances the near-surface temperature. An increase in nighttime temperature dramatically affects the thermal comfort of city inhabitants. As a result, the construction of cool pavements helps reduce the UHI effect on the city (Salamanca et al. 2012; Wang et al. 2016).

The street canyon with both side buildings and road in the middle can be designed with a better height to width ratio. Narrow streets with a lesser sky view factor can result in lower air temperature during daytime (Priyadarsini et al. 2008). However, the design of narrow roads may not be suitable for all types of cities and climatic conditions. Urban canopies due to high-rise buildings and streets reduce the airflow constituting a physical barrier. The design of streets and buildings along the prevailing airflow can significantly help regulate the near-surface air temperature (Priyadarsini et al. 2008). This can be a useful measure for coastal cities experiencing sea breeze.

The design of buildings and streets with respect to the orientation of the sun could affect outdoor thermal comfort (Jamei and Rajagopalan 2019). This measure can have its own modification according to the geographical presence.

The sustainable urban infrastructure can be implemented in a combination of suitable measures depending upon the geographical location and climatic condition. The localized measures like building design, windows design, reflective roof, and wall materials, and urban planning measures like the design of road infrastructure can also be combined for a sustainable UHI mitigation (Kolokotsa et al. 2013; Li et al. 2014).

Sustainable urban infrastructure is generally preferred along with vegetation and greening measures. A large number of studies are focused on the green roof and cool roof/ reflective material technology together and suggested combined adaptation for effective mitigation (Kolokotsa et al. 2013; Li et al. 2014; Li and Norford 2016; Yang et al. 2018). Mohajerani et al. (2017) reviewed the causes, and mitigation of UHI, with reference to the thermal properties of asphalt concrete and suggested the use of measures in combination. A numerical modeling study by Synnefa et al. (2008) on the use of cool material as UHI mitigation found that large-scale increases in albedo could lower ambient air temperatures by 2°C. Similarly, a study on angular reflectance of reflective facades at several inclination angles of solar radiation reveals the behavior and cooling potential of the material as an effective coating over urban paving and building envelope (Rossi et al. 2015). Simulation of the effects of cool roofing and insulation shows that higher albedo material led to superior energy saving and UHI mitigation than insulation measures (Roman et al. 2016). And a proper design of urban forms would largely mitigate UHI especially considering the orientation of buildings in the design plans (Farhadi et al. 2019).

3.3. Water and Soil Management

Sustainable water management implicates using water to meet current, ecological, social, and economic needs optimally. For UHI mitigation, the storm water can be managed to help in evaporative cooling of street-level temperature (Hunt and Collins 2008; Jones et al. 2010). Pavements can be designed to optimize the evaporative cooling effect and reduce direct solar heating. A combination of pervious pavement materials and a water-retentive sub-layer can help absorb water in the subsoil (Nakayama and Fujita 2010; Karasawa et al. 2006). When the upper pavement layer is exposed to direct sun, the absorbed water helps in cooling the ambient air at street level through continuous evaporation. Also, large water bodies inside the cities can reduce the air temperature around them. Stagnant water bodies like lakes, ponds, and fountains can harness the cooling effect of the water body through evaporation. The water can also transport the heat away from the city through rivers and canals. The reverse effect created by water bodies over cities is known as 'Urban Cool Island' (Mostofa and Manteghi 2020).

Very few studies are conducted for UHI mitigation water and soil management measures because of the incomplete definition of cool pavements. The role and influence of cool pavements on air temperature is not so clear (Qin 2015). The mitigation and cooling potential of cool and permeable pavements has the greatest effect and can contribute significantly to decreased temperature in the urban environment, cooling the area by an average of 2°C (Santamouris 2013; Huang and Chen 2020).

3.4. Anthropogenic Heat Reduction

Anthropogenic heat (AH) emissions produced by human activities over the city are one of the contributors of UHI (Li et al. 2013). The sources include heat generated by vehicular emission, industrial processes, heat conduction through building walls or emitted directly into the atmosphere by air-conditioning systems, and the metabolic heat produced by leaving bodies. Particulate matter, water vapor, and carbon dioxide emissions from industrial, residential, and automotive combustion processes enhance air pollution and alter the urban net all-wave radiation budget by altering the amount of incident shortwave solar radiation, long-wave infrared radiation, and ground heat-flux of the urban surface (Asimakopoulos 2001; Emmanuel 2005; Santamouris 2013) and ultimately warming the ambient air to create the UHI. Thus, if the amount of particulate matter can be reduced by effectively filtering the emissions from industries, power plants, sub-stations, implementing strict environmental regulations (EEA 2021), etc. there might be a subsequent decrease in the AH and UHI signatures. Stanhill and Kalma (1995) reported that aerosols could minimize the UHI effect by reducing nearly 33% of incoming shortwave solar radiation. However, sometimes the neutralizing effect of aerosols gets withered away due to the excessive amount of urban air pollution. Several European countries have also begun to promote the growth concept of carbon-negative companies to reduce anthropogenic impacts (EU industrial policy strategy).

The heat generated from commercial spaces and household activities due to air conditioning, cooking, etc. contribute significantly to UHI (Yang et al. 2017). They can be reduced by installing energy-efficient electrical appliances, appropriate arrangement of outdoor units, and systematic use of cooling towers. Improvement of thermal insulation and thermal shielding of buildings with proper heat-insulating materials also helps in reducing the UHI effect (Yamamoto 2006). Moreover, adopting centrally controlled heating or cooling measures also can contribute significantly towards bringing down the level of heat dissipated through anthropogenic sources. As mentioned before, using green roofs and walls can also considerably reduce radiation from buildings (Balany et al. 2020). Additionally, adopting water-retentive materials in construction works is another way to minimize the amount of AH (Nakayama and Fujita 2010). In recent times, efficient traffic control measures, including the use of eco-friendly electric vehicles or low emission vehicles and encouraging the use of bicycles, are the effective actions being taken to deal with such scenarios. Reducing the fossil fuel-based (coal, oil, and gas) power generation can help lessen local air pollution, and promoting increased use of natural renewable and untapped sources of energy can further decrease the AH emission (Karka et al. 2017; Ribeiro et al. 2015) and eventually, it would not encourage the intensification of UHI.

4. Indian Scenario

India is developing at a relatively slower rate compared to China and other advanced countries as far as urbanization is concerned. There are more than 300 Indian cities with a population of over 0.1 million. And there are at least 46 cities with a population of 1 million and above, where eight are considered major ones, i.e., Delhi, Mumbai, Kolkata, Hyderabad, Bengaluru, Chennai

Pune, and Ahmedabad (according to Census of India 2011). Since India is a country that aspires to be recognized as an advanced nation in a few decades from now, the Government of India (GOI) has plans to have 100 smart cities over the years to accommodate the growing aspirations. Although extensive literature is not available for all of the 100 cities included in the smart city project, an effort is made here to review the Indian scenario as far as UHI, consequent effect, and mitigation is concerned.

4.1. UHI over Indian Cities

The increase in the development of heat pockets over major Indian cities is the result of rapid urbanization. Sharma and Bharat (2009) stated that the variation in city climate is observed due to albedo, the change in city structure and shape, pollution, and anthropogenic heat from human activities. In the current scenario, at least 30 Indian cities are experiencing rapid urban growth, including the eight major cities (Figure 2). For instance, urban growth is quite prominent in the case of Hyderabad during the last 15 years, i.e., 2005-2020 (Figure 3). Although the eight major cities are big enough to experience a definite and significant signature of UHI and its effect at a certain time, other growing cities may be experiencing the same in a moderate intensity.

A preliminary analysis is carried out based on the LST values over some of the Indian cities using Landsat-5 and Landsat-8 datasets based on their availability during the summer and winter months. Datasets from a similar period have been chosen for each season of the considered years (Table 1). LST was extracted only for the urban pixels to highlight the spatial variation of SUHI through the surface temperature. Notably, the LST of a given place is influenced by surface air temperature and other surface and sub-surface characteristics such as soil moisture, soil roughness, plant species, elevation, incident shortwave and long-wave radiations, etc. (Prakash et al. 2019). It can be computed for each urban pixel (Giannini et al. 2015; Salih et al. 2004; Mustafa et al. 2020) as:

$$\mathrm{LST} = \frac{\mathrm{BT}}{1+\left(\lambda(\mathrm{BT})\times\frac{\ln(\varepsilon)}{\rho}\right)}$$

where, BT signifies the sensor brightness temperature in Celsius, λ (11.5 μm) is the wavelength of the emitted radiance: $\rho = h(c/\sigma) = 1.438 \times 10^{-2}$ mK, σ is the Stefan–Boltzmann constant, h is Planck's constant, c is the velocity of light, and ε is the land surface emissivity.

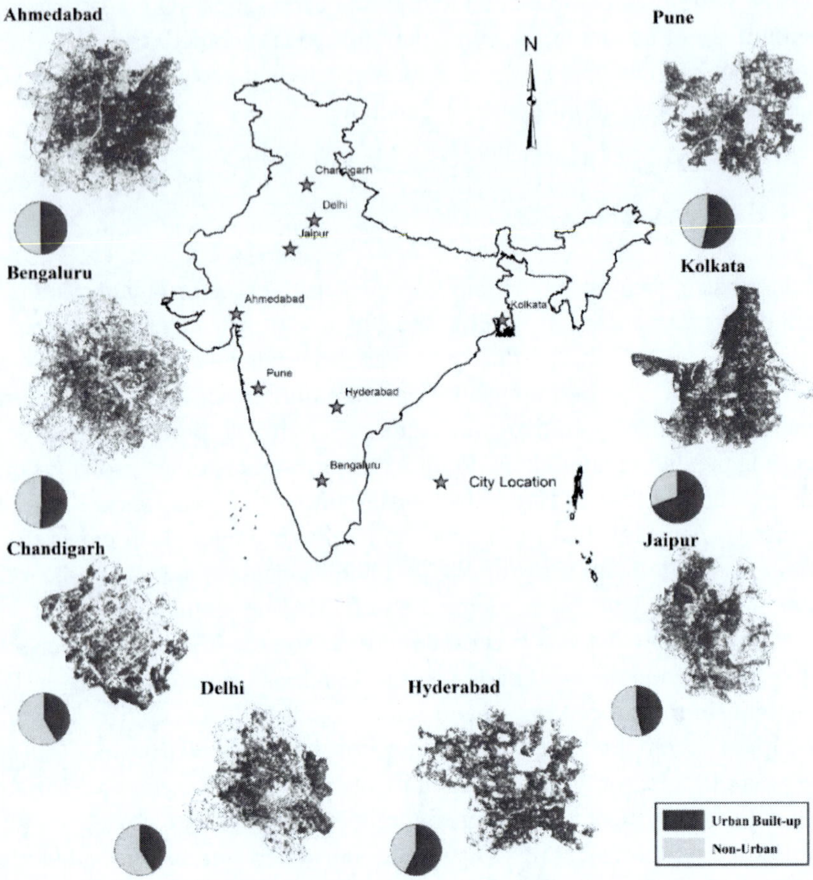

Figure 2. Geospatial location of eight cities of India considered in the preliminary analysis carried out in this study. The urban built-up areas are shown for the year 2020.

Table 1. Maximum, minimum and mean land-surface temperature (°C) over some of the cities of India derived from LANDSAT-5 and LANDSAT-8 observations

City	Year	Season					
		Winter			Summer		
		Minimum	Maximum	Average	Minimum	Maximum	Average
Ahmedabad	2019	18.60	34.81 ↑	25.13 ↑	30.85 ↑	45.84 ↑	38.76 ↑
	1996	18.60	34.66	25.09	27.52	41.85	35.43
Bengaluru	2017	16.00 ↑	31.03 ↑	23.88 ↑	22.53	37.93 ↑	30.95
	1993	12.84	27.52	21.13	23.69	35.27	32.69
Chandigarh	2019	7.02 ↑	16.62 ↑	13.07 ↑	24.90 ↑	45.06 ↑	38.99 ↑
	1998	7.00	16.57	12.77	22.98	36.45	30.53
Delhi	2020	11.34 ↑	25.44 ↑	15.94	26.95 ↑	47.00 ↑	37.99 ↑
	2000	10.93	23.69	17.52	22.82	35.27	29.16
Hyderabad	2020	14.50	32.12 ↑	25.97 ↑	23.66 ↑	40.39 ↑	32.63 ↑
	2005	14.72	31.25	22.80	20.62	39.95	32.59
Kolkata	2019	17.12	24.86	20.20	25.20 ↑	36.78 ↑	30.80 ↑
	1998	19.29	27.52	22.42	21.51	33.68	28.64
Pune	2019	18.49 ↑	32.14 ↑	23.52 ↑	28.74	46.11 ↑	36.60
	1996	18.39	30.02	22.81	30.02	44.85	40.16
Jaipur	2016	17.98 ↑	34.64 ↑	25.71 ↑	27.90 ↑	45.43	37.51
	1998	15.65	25.83	19.75	27.10	45.59	38.01

Hyderabad recorded a mean surface temperature increase from nearly 28.5°C in 2001 to 31.6°C in 2011 and an increase in maximum and minimum temperatures (Sharma 2016). The LST-based analysis (Table 1) observed that during the summer season, the maximum and minimum surface temperature for 2005 was 39.95 and 20.62°C, and for 2020, it was 40.39 and 23.66°C, respectively, indicating the possible impact of urban growth induced SUHI. The (spatial) mean temperature was observed to be 32.59°C for 2005 and 32.63°C for 2020, indicating a marginal increase during summer. In winter, the maximum and minimum surface temperature values for 2005 and 2020 indicate that the underlying surfaces over the cities are possibly getting warmer by ~1°C. Also, the (spatial) mean temperature values indicate that the impact of SUHI is marginal in summer, but it is prominent in winter with an increase of ~3°C. From the spatial variation, it is evident that there has been a substantial increase in surface temperature for both seasons in recent years compared to 2005 (Figure 4). It is also seen that due to the presence of the Hussain Sagar Lake in the central part of the city, the temperatures in the areas adjacent to the lake are quite low as compared to other areas, thus proving that the presence of stagnant water bodies within the city limits can help reduce the impact of UHI and may support in the formation of Urban Cool Island. However, the southern and northern parts of the city are affected by the SUHI during both seasons (Figures 4b, d).

In Kolkata, it was observed that the lowest and highest temperature variation to be around 5.81 and 7.72°C, respectively, from 1990 to 2020 (Halder et al. 2021). On an average, the annual temperature increased by nearly 0.157°C each year between 1990 to 2020 within city limits and the surrounding areas. LST is calculated for the years 1998 and 2019 as per the data availability, and it was observed that there had been a substantial increase in maximum, minimum, and mean surface temperatures in the summer season. However, in the case of winter, LST is observed to decrease, which could be an exceptional scenario associated with possible drop caused by precipitation before the image was acquired and prevailing water bodies in and around the city.

The LST analysis indicates that Bengaluru would experience a SUHI during winter, whereas a "Cool Island Effect" during summer (Table 1). The spatial variability of LST during 2020 shows a "Cool Island Effect" during summer (Figure 5) too. Several studies highlighted such observations (Ramachandra and Kumar 2010; Ambinakudige 2011; Siddiqui et al. 2021) as well. According to these studies, the outskirts (beyond the core of the city) of Bengaluru (Greater Bengaluru) showed an increase of around 2-2.5°C in air temperatures due to excessive urban expansion. However, the "Cool Island Effect" observed in the core city area is possibly due to higher vegetative cover and the presence of water bodies. Pune showed an increase of 1-6°C in LST between the years 1999 and 2006 (Veena et al. 2020). However, the LST-based analysis suggests a similarity with Bengaluru, although the values and years are different (Table 1).

A study conducted by IIT Delhi pointed out that Delhi experiences a maximum UHI intensity of 8.3°C and a minimum of 4.7°C (Mohan et al. 2012). The UHI signature is primarily dominant during the afternoon and night hours over Delhi. Notably, Delhi has shown a consistent warming trend after 1990 in association with the urbanization pattern and consequent UHI effect (Mohan et al. 2011). The LST-based analysis shows an increase in maximum and minimum values during both seasons; however, the spatial mean shows an overall increase during summer only! And the urban response during the summer is quite strong with enhanced UHI, but winter days seem to be affected by smog, resulting in an overall reduction.

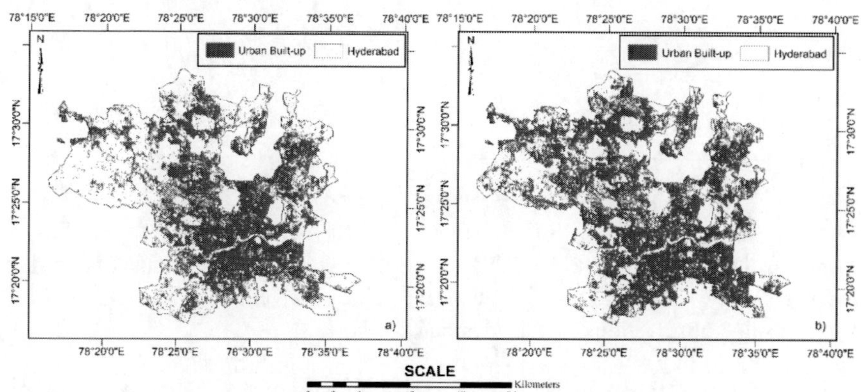

Figure 3. Urban growth over Hyderabad during 2005-2020. The urban area during 2005 is illustrated in (a), and the same is shown in (b) for the year 2020. The urban built-up area is masked out based on the LULC classifications carried out using LANDSAT-5 and LANDSAT-8 observations.

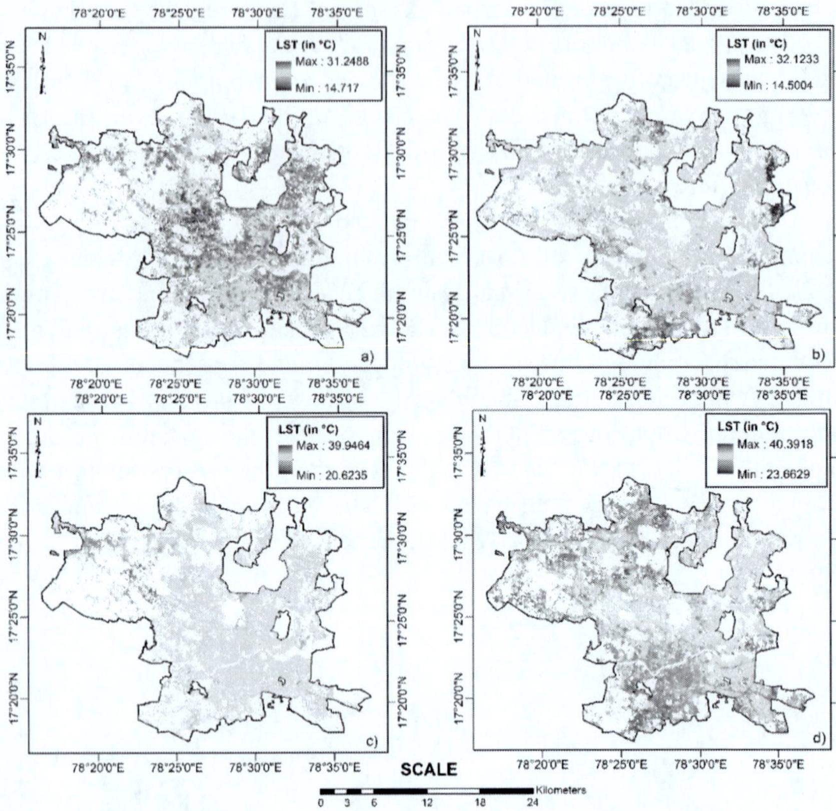

Figure 4. The spatial variation of LST (°C) during winter (a, b) and summer (c, d) of the years 2005 (left panel) and 2020 (right panel) derived from LANDSAT-5 and LANDSAT-8 observations over Hyderabad.

Over Jaipur, the mean maximum UHI intensity of different seasons ranges from 5.12 K to 10.37 K and the overall mean maximum SUHI intensity for all seasons is 7.86 K (Mathew et al. 2017). And the current analysis based on LST indicates similarity with Pune and Bengaluru (Table 1). Therefore, further investigation would be required for drawing any conclusion in this regard.

According to a study by IIT Bhubaneshwar, rapid and unplanned urbanization coupled with LU changes from 2000 and 2014 contributed to an increase of 1.8°C within Bhubaneswar city as compared to the surrounding areas (Swain et al. 2017). The UHI intensity over Ahmedabad increased from around 0.4°C to 0.8°C approximately over a decade (Bajaj et al. 2012). The LST-based analysis (Table 1) suggests the increase (marginal/substantial) of maximum, minimum, and mean values in recent years, both in summer and winter, compared to the late 90s. This is valid for both Ahmedabad and Chandigarh, i.e., a positive SUHI impact is realized over these cities. Over Chennai, the maximum UHI during summer and winter is 2.48 and 3.35°C, respectively (Veena et al. 2020). The study of Maral and Mukhopadhyay (2015) suggested that the core city and suburbs of Mumbai also experience UHI effect. The LST-based analysis by Talukdar et al. (2021) suggests a consistent increase in maximum and minimum temperatures over Delhi and Mumbai due to the contribution from UHI.

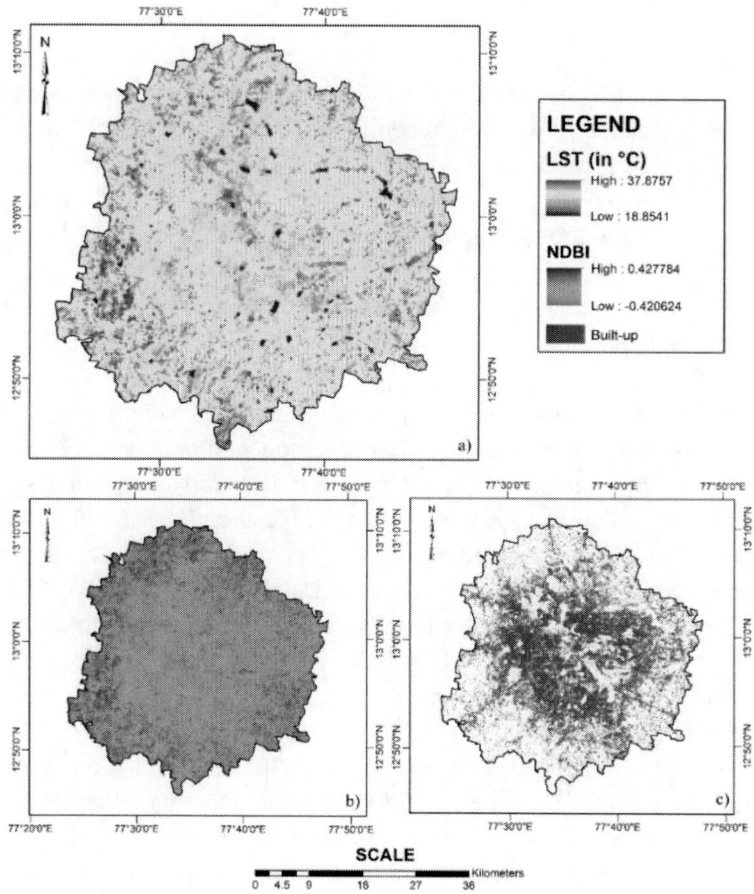

Figure 5. Urbanization over Bengaluru in the year 2020 derived from LANDSAT-8 observations (b, c) along with the spatial variation of LST (a) signifying the Cool Island Effect prevailing in the city.

There are several other studies that analyzed the affirmation of UHI and its effects over several Indian cities such as Vishakhapatnam (Devi 2006), Bhopal (Gupta et al. 2009), Guwahati (Borbora and Das 2014), Kochi (Thomas et al. 2014), Bathinda (Sharma and Pandey 2015), etc. Considering all the studies over India, the maximum UHI intensity observed is 8-9°C while the International maximum is 12°C (Veena et al. 2020).

4.2. Mitigating the UHI Effect in Indian Cities

Cities all over the world are prone to the impact of UHI intensity. The planning and policymakers are implementing mitigation measures discussed in section 3 to cut down the severe impact of UHI. In India, the incorporation of mitigation measures is relatively recent; therefore, very few such applications can be seen. For instance, a group of researchers from Ernest Orlando Lawrence Berkeley National Laboratory and International Institute for Information Technology Hyderabad conducted a pilot study over Hyderabad. This study demonstrated the effectiveness of cool roof technology as one of the UHI control measures and suggested a reduction of ~3.5°C while this technology is adopted aggressively besides using reflective materials (Akbari et al. 2011). A study over Varanasi by Agnihotri et al. (2018) investigated the impact of green spaces on the urban microclimate. The central city area with high-density built-ups experienced higher surface temperatures than the open areas covered by vegetation and water bodies. The study also emphasized the importance of urban greenery for a sustainable urban environment, healthy and quality life. Pramanik and Punia (2019) evaluated the cooling effect of different urban greenery levels at local and park levels for Delhi. Park level analysis was done using park cooling intensity (PCI) and its relationship with urban greenery composition and configuration. The percentage of land cover or the average size of the green patch concerning the urban built-up is found to be significant in reducing the UHI.

A study over Nagpur (Kotharkar et al. 2020) demonstrated a systematic approach for selecting and implementing mitigation strategies suitable for the different local climatic zones (LCZ). It identified highly critical LCZs over Nagpur and performed simulations by considering greening and using reflective surfaces, namely cool roofs and cool pavement. The outcome was mostly similar to the previously reported usefulness of greening and vegetation strategies, specifically over sparsely built and planned settlements with open spaces. The older and unplanned settlements with dense urban agglomeration can adopt the use of reflective surfaces more significantly than other measures. Another study over Nagpur (Jain et al. 2020) also suggested from extensive surveys and analysis that green areas and landscaping should be prioritized as a mitigation measure for improving the localized thermal environment.

A study over the English bazaar town area of West Bengal by Ziaul and Pal (2020) conducted artificial microclimatic simulations to evaluate the performance of green infrastructure-based UHI mitigation measures. The study suggested 100% greening of roof and walls and plantation at suitable sites along with a 50% greening of both roof and wall can be suitable for reducing the air temperature up to ~2.6°C (Ziaul and Pal 2020). Similar results were also obtained by an earlier study by Ghosh and Das (2018) over the Rajarhat block near the Kolkata area. The size and shape of green patches and water bodies impacted the reach and potential of cooling, where irregular shapes of green space and water bodies play an important role in UHI mitigation.

The study by Gupta et al. (2019) analyzed the impact of water bodies like lakes and rivers inside Ahmedabad and Chandigarh on their microclimate. An appreciable average temperature dip of 7.51 and 3.12°C is observed respectively during summer and winter near the Sukhna lake in Chandigarh city. An average dip of about 1.57°C (0.69°C) and 1.71°C (0.65°C) observed during summer and winter, respectively on the right (left) bank of the Sabarmati river in Ahmedabad.

Some of the studies over Indian cities focused on urban infrastructure modification too. Kanji et al. (2015) attempted to analyze the UHI phenomenon over Roorkee and established the relationship between architectural styles and climatic differences. The study suggests the adaptation of indigenous design architecture as an effective method to mitigate UHI impacts. A study by Kandya and Mohan (2018) investigated the thermal performance of composite building materials like 'Bamboo-Concrete', 'Rammed Earth', and the impact of building envelope modifications like increased wall thickness and cavity wall through simulation. They suggested a 7.5% reduction in power consumption for space cooling using 'Bamboo-Concrete', and 4.7-5.3% using increased wall thickness, could be adopted for mitigating the UHI effect. Thomas et al. (2018) reported the nanoparticles-based coating possibilities as an effective method to mitigate UHI over Kochi. The nanoparticles can be used effectively as an additional coating to provide higher albedo, reducing building materials' storage heat and surface temperature.

Conclusion and Way Forward

UHI is a susceptible phenomenon that governs many aspects of a well-developed and industrially rich city, although the Cool Island Effect is possible in exceptional situations. The urbanization induced UHI can impact the localized and mesoscale weather, microclimate, comfort level and health of urban dwellers, air and water quality, energy consumption, etc. Therefore, appropriate mitigation measures must be taken irrespective of the socio-economic conditions. The mitigation measures include developing and adopting the cool-roof technology, using appropriate materials for sustainable urban infrastructure, managing the vegetation cover and adopting green technology, reducing the AH and greenhouse gas emission, proper soil and water management, etc. However, a sincere effort to study the scientific and technological aspects associated with these is required in each country aspiring to be an advanced economy. It would help in understanding the UHI effect specific to a particular city and aid in developing technology for mitigation purposes.

Although advanced nations like USA or countries in Europe or some in Asia have already been able to equip themselves with scientific and technological aspects relating to UHI and its mitigation, India still lags behind. For instance, appropriate and adequate satellite observations are not freely available for academic and research use to study the UHI and its effects. Besides, India lacks in establishing a dense mesoscale observational network for city-based studies. Also, numerical modeling could be quite handy for the studies relating to UHIs and help predict the future city climate. Application of computational fluid dynamics (CFD) could be useful in urban microclimate studies (Li and Norford 2016; Yuan et al. 2020), but India stands in its preliminary stage in this sense. The application of mesoscale and CFD models for the study of the UHI effect over Indian cities is very rare. Even to date, the applications in urban infrastructures are also not reflected widely in India.

Therefore, India lacks both observational and numerical modeling studies for the scientific research of UHI and its effects. Also, there are fewer efforts to develop the in-house technology (e.g., urban data-base development, green technology for urban infrastructure) required to mitigate UHI effects and effective policies for adopting the relevant measures so that the same is reflected more in the urban infrastructures. Thus, sincere, significant, and active efforts are needed to acquire the required scientific knowledge and technological expertise in the future. Since the lack of community education and outreach could be a possible reason for the increased UHI intensity

(Sharma and Bharat 2009), policies should be framed and implemented for maintaining a balance in the microclimate of cities by communicating the same to the public.

References

Agnihotri, A. K. (2018). Impact of green spaces on the urban microclimate through Landsat 8 and TIRS Data, in Varanasi, India. *International Journal of Environment and Sustainability,* 7(2).

Akbari, H., Xu, T., Taha, H., Wray, C., Sathaye, J., Garg, V., ... & Reddy, K. N. (2011). Using cool roofs to reduce energy use, greenhouse gas emissions, and urban heat-island effects: Findings from an India experiment (No. LBNL-4746E). *Lawrence Berkeley National Lab.(LBNL)*, Berkeley, CA (United States).

Ambinakudige, S. (2011). Remote sensing of land cover's effect on surface temperatures: a case study of the urban heat island in Bangalore, India. *Applied GIS.*, 7(1).

ARAbI, R., SHAHIdAn, M. F., Kamal, M. M., Jaafar, M. F. Z. B., & Rakhshandehroo, M. (2015). Mitigating urban heat island through green roofs. *Current World Environment.*, 10(1), 918-927.

Argüeso, D., Evans, J. P., Pitman, A. J., & Di Luca, A. (2015). Effects of city expansion on heat stress under climate change conditions. *PLoS one*, 10(2), e0117066.

Avissar, R. (1995). Recent advances in the representation of land-atmosphere interactions in general circulation models. *Reviews of Geophysics.*, 33(S2), 1005-1010.

Bajaj, D. N., Inamdar, A. B., & Vaibhav, V. (2012). Temporal variation of Urban Heat Island using landsat data: a case study of Ahmedabad, India. In 33rd Asian Conference on Remote Sensing 2012, ACRS-2012 (pp. 797-804). *Asian Association on Remote Sensing.*

Balany, F., Ng, A. W., Muttil, N., Muthukumaran, S., & Wong, M. S. (2020). Green infrastructure as an urban heat island mitigation strategy—a review. *Water.*, 12(12), 3577.

Baniya, B., Techato, K. A., Ghimire, S. K., & Chhipi-Shrestha, G. (2018). A review of green roofs to mitigate urban heat island and Kathmandu valley in Nepal. *Appl. Ecol. Environ. Sci*, 6(4), 137-152.

Barros, A. P., & Hwu, W. (2002). A study of land-atmosphere interactions during summertime rainfall using a mesoscale model. *Journal of Geophysical Research: Atmospheres.*, 107(D14), ACL-17.

Battista, G., Evangelisti, L., Guattari, C., De Lieto Vollaro, E., De Lieto Vollaro, R., & Asdrubali, F. (2020). Urban heat island mitigation strategies: experimental and numerical analysis of a university campus in Rome (Italy). *Sustainability.*, 12(19), 7971.

Block, A. H., Livesley, S. J., & Williams, N. S. (2012). *Responding to the urban heat island: a review of the potential of green infrastructure*. Available at https://apo.org.au/sites/default/files/resource-files/2012-03/apo-nid237206.pdf

Bornstein, R. D. (1968). Observations of the urban heat island effect in New York City. *Journal of Applied Meteorology and Climatology*., 7(4), 575-582.

Bornstein, R., & LeRoy, M. (1990). Urban barrier effects on convective and frontal thunderstorms. *In Extended Abstracts, Fourth Conf. on Mesoscale Processes* (pp. 120-121).

Brohan, P., Kennedy, J. J., Harris, I., Tett, S. F., & Jones, P. D. (2006). Uncertainty estimates in regional and global observed temperature changes: A new data set from 1850. *Journal of Geophysical Research: Atmospheres*., 111(D12).

Chander G., & Markham B. (2003). Revised Landsat-5 TM radiometric calibration procedures and postcalibration dynamic ranges. *IEEE Transactions on Geoscience and Remote Sensing*., 41, 2674-2677.

Changnon, S. A. (1981). METROMEX: A review and summary, Meteor. Monogr. No. 40. *Amer. Meteor. Soc.*

Chapman, S., Watson, J. E., Salazar, A., Thatcher, M., & McAlpine, C. A. (2017). The impact of urbanization and climate change on urban temperatures: a systematic review. *Landscape Ecology*., 32(10), 1921-1935.

Chun, B., & Guldmann, J. M. (2018). Impact of greening on the urban heat island: Seasonal variations and mitigation strategies. *Computers, Environment and Urban Systems*., 71, 165-176.

Comarazamy, D. E., González, J. E., Luvall, J. C., Rickman, D. L., & Mulero, P. J. (2010). A land–atmospheric interaction study in the coastal tropical city of San Juan, Puerto Rico. *Earth Interactions*., 14(16), 1-24.

Devi, S. S. (2006). Urban heat islands and environmental impact. In *86th American meteorological society annual meeting*. Atlanta, Georgia.

Dixon, P. G., & Mote, T. L. (2003). Patterns and causes of Atlanta's urban heat island–initiated precipitation. *Journal of Applied Meteorology*., 42(9), 1273-1284.

Dwivedi, A., & Mohan, B. K. (2018). Impact of green roof on micro climate to reduce Urban Heat Island. *Remote Sensing Applications: Society and Environment*., 10, 56-69.

Emmanuel, R.. (2005). *An Urban Approach To Climate Sensitive Design: Strategies for the Tropics.* Available at https://doi.org/10.4324/9780203414644

EPA. (2008). Reducing Urban Heat Islands: Compendium of Strategies. Retrieved March 24, 2013. Available at http://www.epa.gov/heatisland/resources/pdf/GreenRoofs Compendium.pdf

Falasca, S., Ciancio, V., Salata, F., Golasi, I., Rosso, F., & Curci, G. (2019). High albedo materials to counteract heat waves in cities: An assessment of meteorology, buildings energy needs and pedestrian thermal comfort. *Building and environment*., 163, 106242.

Farhadi, H., Faizi, M., & Sanaieian, H. (2019). Mitigating the urban heat island in a residential area in Tehran: Investigating the role of vegetation, materials, and orientation of buildings. *Sustainable Cities and Society*., 46, 101448.

Feyisa, G. L., Dons, K., & Meilby, H. (2014). Efficiency of parks in mitigating urban heat island effect: An example from Addis Ababa. *Landscape and urban planning*., 123, 87-95.

Gamble, J. L., Hurley, B. J., Schultz, P. A., Jaglom, W. S., Krishnan, N., & Harris, M. (2013). Climate change and older Americans: state of the science. *Environmental health perspectives.*, 121(1), 15-22.

Ghosh, S., & Das, A. (2018). Modelling urban cooling island impact of green space and water bodies on surface urban heat island in a continuously developing urban area. *Modeling Earth Systems and Environment.*, 4(2), 501-515.

Giannini, M. B., Belfiore, O. R., Parente, C., & Santamaria, R. (2015). Land Surface Temperature from Landsat 5 TM images: comparison of different methods using airborne thermal data. *Journal of Engineering Science & Technology Review.*, 8(3).

Gibson, J., Boe-Gibson, G., & Stichbury, G. (2015). Urban land expansion in India 1992–2012. *Food Policy.*, 56, 100-113.

Giguère, M. (2009). Literature review of Urban Heat Island mitigation strategies. *Direction de la santé environnementale et de la toxicologie, Institut national de santé publique Québec.* Available at https://numerique.banq.qc.ca/patrimoine/details/52327/2220185

Gunawardena, K. R., Wells, M. J., & Kershaw, T. (2017). Utilising green and bluespace to mitigate urban heat island intensity. *Science of the Total Environment.*, 584, 1040-1055.

Gupta, A., Dey, V., & Goel, A. (2009). Mitigation of thermal pollution to enhance urban air quality through remote sensing and GIS. *10th ESRI India User Conf Noida (NCR).* Available at https://www.esri.in/~/media/esri-india/files/pdfs/events/uc2009/papers/URB1.pdf

Gupta, N., Mathew, A., & Khandelwal, S. (2019). Analysis of cooling effect of water bodies on land surface temperature in nearby region: A case study of Ahmedabad and Chandigarh cities in India. *The Egyptian Journal of Remote Sensing and Space Science.*, 22(1), 81-93.

Halder, B., Bandyopadhyay, J., & Banik, P. (2021). Monitoring the effect of urban development on urban heat island based on remote sensing and geo-spatial approach in Kolkata and adjacent areas, India. *Sustainable Cities and Society*, 74, 103186.

Hansen, J., Ruedy, R., Sato, M., & Lo, K. (2010). Global surface temperature change. *Reviews of Geophysics*, 48(4).

Hegazy, I. R., & Kaloop, M. R. (2015). Monitoring urban growth and land use change detection with GIS and remote sensing techniques in Daqahlia governorate Egypt. *International Journal of Sustainable Built Environment*, 4(1), 117-124.

Herath, H. M. P. I. K., Halwatura, R. U., & Jayasinghe, G. Y. (2018). Evaluation of green infrastructure effects on tropical Sri Lankan urban context as an urban heat island adaptation strategy. *Urban Forestry & Urban Greening*, 29, 212-222.

Huang, J. M., & Chen, L. C. (2020). A numerical study on mitigation strategies of urban heat islands in a tropical megacity: A case study in Kaohsiung City, Taiwan. *Sustainability*, 12(10), 3952.

Huang, Q., & Lu, Y. (2015). The effect of urban heat island on climate warming in the Yangtze River Delta urban agglomeration in China. *International Journal of Environmental Research And Public Health*, 12(8), 8773-8789.

Huff, F. A., & Changnon Jr, S. A. (1973). Precipitation modification by major urban areas. *Bulletin of the American Meteorological Society*, 54(12), 1220-1233.

Hunt, W. F., & Collins, K. A. (2008). Permeable pavement: Research update and design implications. North Carolina Cooperative Extension. Available at https://nacto.org/docs/usdg/urban_waterways_permeable_pavement_hunt.pdf

INDIA, P. (2011). Census of India 2011 provisional population totals. New Delhi: Office of the Registrar General and Census Commissioner. Available at https://censusindia.gov.in/2011-prov-results/paper2/data_files/mp/4-position-iv.pdf

IPCC (2014) Climate change 2014: impacts, adaptation, and vulnerability. Part A: global and sectoral aspects. Contribution of working group II to the fifth assessment report of the intergovernmental panel on climate change [Field, C.B., V.R. Barros, D.J. Dokken, K.J. Mach, M.D. Mastrandrea, T.E. Bilir, M. Chatterjee, K.L. Ebi, Y.O. Estrada, R.C. Genova, B. Girma, E.S. Kissel, A.N. Levy, S. MacCracken, P.R. Mastrandrea, and L.L. White (eds.)]. Cambridge University Press, Cambridge, United Kingdom and New York, NY, USA.

Jain, S., Sannigrahi, S., Sen, S., Bhatt, S., Chakraborti, S., & Rahmat, S. (2020). Urban heat island intensity and its mitigation strategies in the fast-growing urban area. *Journal of Urban Management*, 9(1), 54-66.

Jamei, E., & Rajagopalan, P. (2019). Effect of street design on pedestrian thermal comfort. *Architectural Science Review*, 62(2), 92-111.

Jones, D., Harvey, J., Wang, T., Wu, R., & Campbell, B. (2010). Laboratory testing and modeling for structural performance of fully permeable pavements. Available at https://escholarship.org/uc/item/784252zg

Kandya, A., & Mohan, M. (2018). Mitigating the Urban Heat Island effect through building envelope modifications. *Energy and Buildings*, 164, 266-277.

Kanji, R., Sahu, S., & Agarwal, R. (2015). A Study of the Effect of Building Designs on Climate Change: Mitigation and Adaption. *International Journal of Engineering and Technology*, 7(2), 145.

Karasawa, A., Toriiminami, K., Ezumi, N., & Kamaya, K. (2006, November). Evaluation of performance of water-retentive concrete block pavements. In 8th International conference on concrete block paving.

Karka, P., Papadokonstantakis, S., & Kokossis, A. (2017). Cradle-to-gate assessment of environmental impacts for a broad set of biomass-to-product process chains. The *International Journal of Life Cycle Assessment*, 22(9), 1418-1440.

Kim, H., Gu, D., & Kim, H. Y. (2018). Effects of Urban Heat Island mitigation in various climate zones in the United States. *Sustainable Cities and Society*, 41, 841-852.

Kolokotsa, D., Santamouris, M., & Zerefos, S. C. (2013). Green and cool roofs' urban heat island mitigation potential in European climates for office buildings under free floating conditions. *Solar Energy*, 95, 118-130.

Kotharkar, R., Bagade, A., & Singh, P. R. (2020). A systematic approach for urban heat island mitigation strategies in critical local climate zones of an Indian city. *Urban Climate*, 34, 100701.

Landsberg, H. E., 1956: The climate of towns. Man's Role in Changing the Face of the Earth, W. L. Thomas, Ed., University of Chicago Press, 584–603.

Lead, F. C., Zamuda, C. D., Bilello, D. E., Conzelmann, G., Mecray, E., Satsangi, A., Tidwell, V., Walker, B. J., Pryor, S. C. and Zamuda, C., Energy Supply, Delivery, and Demand. Available at https://nca2018.globalchange.gov/downloads/NCA4_Ch04_Energy_Full.pdf

Leal Filho, W., Echevarria Icaza, L., Emanche, V. O., & Quasem Al-Amin, A. (2017). An evidence-based review of impacts, strategies and tools to mitigate urban heat islands. *International Journal of Environmental Research and Public Health*, 14(12), 1600.

Leal Filho, W., Icaza, L. E., Neht, A., Klavins, M., & Morgan, E. A. (2018). Coping with the impacts of urban heat islands. A literature based study on understanding urban heat vulnerability and the need for resilience in cities in a global climate change context. *Journal of Cleaner Production*, 171, 1140-1149.

Leal Filho, W., Wolf, F., Castro-Díaz, R., Li, C., Ojeh, V. N., Gutiérrez, N., ... & Bönecke, J. (2021). Addressing the urban heat islands effect: A cross-country assessment of the role of green infrastructure. *Sustainability*, 13(2), 753.

Li, D., Bou-Zeid, E., & Oppenheimer, M. (2014). The effectiveness of cool and green roofs as urban heat island mitigation strategies. *Environmental Research Letters*, 9(5), 055002.

Li, X. X., & Norford, L. K. (2016). Evaluation of cool roof and vegetations in mitigating urban heat island in a tropical city, Singapore. *Urban Climate*, 16, 59-74.

Li, X. X., Koh, T. Y., Entekhabi, D., Roth, M., Panda, J., & Norford, L. K. (2013). A multi-resolution ensemble study of a tropical urban environment and its interactions with the background regional atmosphere. *Journal of Geophysical Research: Atmospheres*, 118(17), 9804-9818.

Li, X., Zhou, Y., Asrar, G. R., & Zhu, Z. (2018). Developing a 1 km resolution daily air temperature dataset for urban and surrounding areas in the conterminous United States. *Remote Sensing of Environment*, 215, 74-84.

Li, X., Zhou, Y., Yu, S., Jia, G., Li, H., & Li, W. (2019). Urban heat island impacts on building energy consumption: A review of approaches and findings. *Energy*, 174, 407-419.

Li, Z. L., Tang, B. H., Wu, H., Ren, H., Yan, G., Wan, Z., ... & Sobrino, J. A. (2013). Satellite-derived land surface temperature: Current status and perspectives. *Remote Sensing of Environment*, 131, 14-37.

Liang, X., Xie, Z., & Huang, M. (2003). A new parameterization for surface and groundwater interactions and its impact on water budgets with the variable infiltration capacity (VIC) land surface model. *Journal of Geophysical Research: Atmospheres*, 108(D16).

Lin, C. Y., Chen, W. C., Chang, P. L., & Sheng, Y. F. (2011). Impact of the urban heat island effect on precipitation over a complex geographic environment in northern Taiwan. *Journal of Applied Meteorology and Climatology*, 50(2), 339-353.

Maral, S. G., & Mukhopadhyay, T. (2015). Signal of urban heat island (UHI) effect: a case study of Mumbai metropolitan region. *Mausam*, 66(4), 729-740.

Mather, G. K. (1991). Coalescence enhancement in large multicell storms caused by the emissions from a Kraft paper mill. *Journal of Applied Meteorology and Climatology*, 30(8), 1134-1146.

Mathew, A., Khandelwal, S., & Kaul, N. (2017). Investigating spatial and seasonal variations of urban heat island effect over Jaipur city and its relationship with vegetation, urbanization and elevation parameters. *Sustainable Cities and Society*, 35, 157-177.

Maxwell, K., Grambsch, A., Kosmal, A., Larson, L., & Sonti, N. (2018). Built environment, urban systems, and cities. In: Reidmiller, DR; Avery, CW; Easterling, DR; Kunkel, KE; Lewis, KLM; Maycock, TK; Stewart, BC, eds. 2018. *Impacts, Risks, and Adaptation in the United States: Fourth National Climate Assessment*, Volume II. Washington, DC: US Global Change Research Program. pp. 438–478., 2, 438-478.

Mirzaei, P. A., & Haghighat, F. (2010). Approaches to study urban heat island–abilities and limitations. *Building and Environment*, 45(10), 2192-2201.

Mohajerani, A., Bakaric, J., & Jeffrey-Bailey, T. (2017). The urban heat island effect, its causes, and mitigation, with reference to the thermal properties of asphalt concrete. *Journal of Environmental Management*, 197, 522-538.

Mohan, M., Kandya, A., & Battiprolu, A. (2011). Urban heat island effect over national capital region of India: a study using the temperature trends. *Journal of Environmental Protection*, 2(04), 465.

Mohan, M., Kikegawa, Y., Gurjar, B. R., Bhati, S., Kandya, A., & Ogawa, K. (2012). Urban heat island assessment for a tropical urban airshed in India. Available at https://doi.org/10.4236/acs.2012.22014

Mostofa, T., & Manteghi, G. (2020). Influential factors of water body to enhance the urban cooling islands (Ucis): A review. International Transaction *Journal of Engineering, Management, & Applied Sciences & Technologies*, 11(2), 1-12.

Mustafa, E. K., Co, Y., Liu, G., Kaloop, M. R., Beshr, A. A., Zarzoura, F., & Sadek, M. (2020). Study for predicting land surface temperature (LST) using landsat data: a comparison of four algorithms. *Advances in Civil Engineering*, 2020.

Nakayama, T., & Fujita, T. (2010). Cooling effect of water-holding pavements made of new materials on water and heat budgets in urban areas. *Landscape and Urban Planning*, 96(2), 57-67.

Niyogi, D., Mahmood, R., & Adegoke, J. O. (2009). Land-use/land-cover change and its impacts on weather and climate. *Boundary Layer Meteorology*, 133(3), 297.

Niyogi, D., Pyle, P., Lei, M., et al., 2011. Urban modification of thunderstorms: an observational storm climatology and model case study for the Indianapolis urban region. *Journal of Applied Meteorology and Climatology*, 50, 1129-1144.

Oke, T. R. (1995). The heat island of the urban boundary layer: characteristics, causes and effects. In Wind climate in cities (pp. 81-107). Springer, Dordrecht.

Onishi, A., Cao, X., Ito, T., Shi, F., & Imura, H. (2010). Evaluating the potential for urban heat-island mitigation by greening parking lots. *Urban Forestry & Urban Greening*, 9(4), 323-332.

Perera, E. M., & Sanford, T. (2011). Rising Temperatures, Worsening Ozone Pollution. Union of Concerned Scientists. Available at https://www.ucsusa.org/sites/default/files/2019-09/climate-change-and-ozone-pollution.pdf

Peterson, T. C., Gallo, K. P., Lawrimore, J., Owen, T. W., Huang, A., & McKittrick, D. A. (1999). Global rural temperature trends. *Geophysical Research Letters*, 26(3), 329-332.

Prakash, S., Shati, F., Norouzi, H., & Blake, R. (2019) Observed differences between near-surface air and skin temperatures using satellite and ground-based data. *Theoretical and Applied Climatology*, 137, 587–600.

Pramanik, S., & Punia, M. (2019). Assessment of green space cooling effects in dense urban landscape: a case study of Delhi, India. *Modeling Earth Systems and Environment*, 5(3), 867-884.

Price, A., Jones, E. C., & Jefferson, F. (2015). Vertical greenery systems as a strategy in urban heat island mitigation. *Water, Air, & Soil Pollution*, 226(8), 1-11.

Priyadarsini, R., Hien, W. N., & David, C. K. W. (2008). Microclimatic modeling of the urban thermal environment of Singapore to mitigate urban heat island. *Solar energy*, 82(8), 727-745.

Qin, Y. (2015). A review on the development of cool pavements to mitigate urban heat island effect. *Renewable and Sustainable Energy Reviews*, 52, 445-459.

Ramachandra, T. V., & Kumar, U. (2010). Greater Bangalore: Emerging urban heat island. GIS development, 14(1), 86-104.

Ribeiro, I., Monteiro, A., Martins, H., Freitas, S., Borrego, C., & Lopes, M. (2015). How does the use of biodiesel affect urban air quality?. *International Journal of Environment and Pollution*, 58(1-2), 79-88.

Roman, K. K., O'Brien, T., Alvey, J. B., & Woo, O. (2016). Simulating the effects of cool roof and PCM (phase change materials) based roof to mitigate UHI (urban heat island) in prominent US cities. *Energy*, 96, 103-117.

Rosenfeld, D. (2000). Suppression of rain and snow by urban and industrial air pollution. *Science*, 287(5459), 1793-1796.

Rosenzweig, C., Solecki, W., & Slosberg, R. (2006). Mitigating New York City's heat island with urban forestry, living roofs, and light surfaces. A report to the New York State Energy Research and Development Authority. Available at http://citeseerx.ist.psu.edu/viewdoc/download?doi=10.1.1.543.4848&rep=rep1&type=pdf

Rossi, F., Castellani, B., Presciutti, A., Morini, E., Filipponi, M., Nicolini, A., & Santamouris, M. (2015). Retroreflective façades for urban heat island mitigation: Experimental investigation and energy evaluations. *Applied Energy*, 145, 8-20.

Sachindra, D. A., Ng, A. W. M., Muthukumaran, S., & Perera, B. J. C. (2016). Impact of climate change on urban heat island effect and extreme temperatures: a case-study. *Quarterly Journal of the Royal Meteorological Society*, 142(694), 172-186.

Sailor, D. J. (2013). Energy buildings and urban environment. In *Climate Vulnerability: Understanding and Addressing Threats to Essential Resources* (pp. 167-182). Elsevier Inc.

Salamanca, F., Martilli, A., & Yagüe, C. (2012). A numerical study of the Urban Heat Island over Madrid during the DESIREX (2008) campaign with WRF and an evaluation of simple mitigation strategies. *International Journal of Climatology*, 32(15), 2372-2386.

Salih, M. M., Jasim, O. Z., Hassoon, K. I., & Abdalkadhum, A. J. (2018). Land surface temperature retrieval from LANDSAT-8 thermal infrared sensor data and validation with infrared thermometer camera. *International Journal of Engineering & Technology*, 7(4.20), 608-612.

Santamouris, M. (2013). Energy and climate in the urban built environment. Routledge. Available at https://www.taylorfrancis.com/books/mono/10.4324/9781315073774/energy-climate-urban-built-environment-santamouris

Santamouris, M. (2013). Using cool pavements as a mitigation strategy to fight urban heat island—A review of the actual developments. *Renewable and Sustainable Energy Reviews*, 26, 224-240.

Santamouris, M., & Asimakopoulos, D. N. (2001). Energy and climate in the urban built environment. London: James X James. Available at https://doi.org/10.4324/9781315073774

Sharma, D., & Bharat, A. (2009). Urban Heat Island Effect-Causes, Impacts, Methods of Measurement and Mitigation Options. *Institute of Town Planners India*, *6*(2), 69-77.

Sharma, N., & Pandey, P. (2015). Study of urban heat island in Bathinda city, Punjab. In *16th ESRI India User Conference* (pp. 1-14).

Sharma, V. (2016). Green infrastructure measures to mitigate urban heat island effect: a case of Hyderabad city (Doctoral dissertation, SPA Bhopal).

Shepherd, J. M. (2005). A review of current investigations of urban-induced rainfall and recommendations for the future. *Earth Interactions*, 9(12), 1-27.

Shepherd, J. M., & Burian, S. J. (2003). Detection of urban-induced rainfall anomalies in a major coastal city. *Earth Interactions*, 7(4), 1-17.

Siddiqui, A., Kushwaha, G., Raoof, A., Verma, P. A., & Kant, Y. (2021). Bangalore: urban heating or urban cooling?. The Egyptian *Journal of Remote Sensing and Space Science*, 24(2), 265-272.

Singh, N., Mhawish, A., Ghosh, S., Banerjee, T., & Mall, R. K. (2019). Attributing mortality from temperature extremes: A time series analysis in Varanasi, India. *Science of the Total Environment*, 665, 453-464.

Somers, K. A., Bernhardt, E. S., McGlynn, B. L., & Urban, D. L. (2016). Downstream Dissipation of storm flow heat pulses: A case study and its landscape-level implications. *JAWRA Journal of the American Water Resources Association*, 52(2), 281-297.

Stanhill, G., & Kalma, J. D. (1995). Solar dimming and urban heating at Hong Kong. *International Journal of Climatology*, 15(8), 933-941.

Sun, R., & Chen, L. (2017). Effects of green space dynamics on urban heat islands: Mitigation and diversification. *Ecosystem Services*, 23, 38-46.

Swain, D., Roberts, G. J., Dash, J., Lekshmi, K., Vinoj, V., & Tripathy, S. (2017). Impact of rapid urbanization on the city of Bhubaneswar, India. *Proceedings of the National Academy of Sciences, India Section A: Physical Sciences*, 87(4), 845-853.

Synnefa, A., Dandou, A., Santamouris, M., Tombrou, M., & Soulakellis, N. (2008). On the use of cool materials as a heat island mitigation strategy. *Journal of Applied Meteorology and Climatology*, 47(11), 2846-2856.

Takebayashi, H., & Moriyama, M. (2009). Study on the urban heat island mitigation effect achieved by converting to grass-covered parking. *Solar Energy*, 83(8), 1211-1223.

Talukdar, S., Rihan, M., Hang, H. T., Bhaskaran, S., & Rahman, A. (2021). Modelling urban heat island (UHI) and thermal field variation and their relationship with land use indices over Delhi and Mumbai metro cities. *Environment, Development and Sustainability*, 1-29.

Thomas, G., Sherin, A. P., Ansar, S., & Zachariah, E. J. (2014). Analysis of urban heat island in Kochi, India, using a modified local climate zone classification. *Procedia Environmental Sciences*, 21, 3-13.

Thorne, K., Markham, B., Barker, P. S., & Biggar, S. J. P. E. (1997). Radiometric calibration of Landsat. *Photogrammetric Engineering & Remote Sensing, 63*(7), 853-858.

Tuholske, C., Caylor, K., Funk, C., Verdin, A., Sweeney, S., Grace, K., ... & Evans, T. (2021). Global urban population exposure to extreme heat. *Proceedings of the National Academy of Sciences*, 118(41).

Tzavali, A., Paravantis, J. P., Mihalakakou, G., Fotiadi, A., & Stigka, E. (2015). Urban heat island intensity: A literature review. *Fresenius Environmental Bulletin*, 24(12b), 4537-4554.

U.S. Climate Change Science Program. 2008. Analyses of the effects of global change on human health and welfare and human systems. A Report by the U.S. Climate Change Science Program and the Subcommittee on Global Change Research. [Gamble, J.L. (ed.), K.L. Ebi, F.G. Sussman, T.J. Wilbanks (Authors)]. U.S. Environmental Protection Agency, Washington, DC, USA.

UN. (2014). World Urbanization Prospects: The 2014 Revision-Highlights. UN. Available at https://www.un.org/en/development/desa/publications/2014-revision-world-urbanization-prospects.html#:~:text=The%202014%20revision%20of%20the, population%20between%202014%20and%202050.

UNFPA. Growing up urban: state of world population 2007—youth supplement. New York, NY: United Nations Population Fund, 2007. Available at https://sustainable development.un.org/index.php?page=view&type=400&nr=1109&menu=35

USGS, Landsat 8 Data Users Handbook, USGS, Reston, VA, USA, 2016. Available at https://www.usgs.gov/media/files/landsat-8-data-users-handbook

Veena, K., Parammasivam, K. M., & Venkatesh, T. N. (2020). Urban Heat Island studies: Current status in India and a comparison with the International studies. *Journal of Earth System Science*, 129(1), 1-15.

Venter, Z. S., Krog, N. H., & Barton, D. N. (2020). Linking green infrastructure to urban heat and human health risk mitigation in Oslo, Norway. *Science of the Total Environment*, 709, 136193.

Voogt, J. A. (2004). Urban heat islands: hotter cities. America Institute of Biological Sciences. Available at www.actionbioscience.org/environment/voogt.html

Voogt, J. A., & Oke, T. R. (2003). Thermal remote sensing of urban climates. *Remote Sensing of Environment*, 86(3), 370-384.

Wang, Y., Berardi, U., & Akbari, H. (2016). Comparing the effects of urban heat island mitigation strategies for Toronto, Canada. *Energy and Buildings*, 114, 2-19.

Watts, N., Amann, M., Arnell, N., Ayeb-Karlsson, S., Belesova, K., Boykoff, M.; Byass, P., Cai, W., Campbell-Lendrum, D., Capstick, S., et al. The 2019 report of The Lancet Countdown on health and climate change: Ensuring that the health of a child born today is not defined by a changing climate. *Lancet*, 2019, 394, 1836–1878.

Xiao, R., Ouyang, Z., Zheng, H., Li, W., Schienke, E. W., & Wang, X., (2007). Spatial pattern of impervious surfaces and their impacts on land surface temperature in Beijing, China. *Journal of Environmental Sciences*, 19, 250-256.

Yun-hao, C. H. E. N., Jie, W., & Xiao-bing, L. I. (2002). A study on urban thermal field in summer based on satellite remote sensing. *Remote Sensing for Land & Resources*, *14*(4), 55-59.

Yamamoto, Y. (2006). Measures to mitigate urban heat islands. NISTEP Science & Technology Foresight Center. Available at https://www.coolrooftoolkit.org/wp-content/uploads/2012/04/Measures-to-Mitigate-UHI-Yamamoto.pdf

Yang, J., Pyrgou, A., Chong, A., Santamouris, M., Kolokotsa, D., & Lee, S. E. (2018). Green and cool roofs' urban heat island mitigation potential in tropical climate. *Solar Energy*, 173, 597-609.

Yang, J., Tham, K. W., Lee, S. E., Santamouris, M., Sekhar, C., & Cheong, D. K. W. (2017). Anthropogenic heat reduction through retrofitting strategies of campus buildings. *Energy and Buildings*, 152, 813-822.

Yuan, C., Adelia, A. S., Mei, S., He, W., Li, X. X., & Norford, L. (2020). Mitigating intensity of urban heat island by better understanding on urban morphology and anthropogenic heat dispersion. *Building and Environment*, 176, 106876.

Zhang, D. L., Shou, Y. X., Dickerson, R. R., & Chen, F. (2011). Impact of upstream urbanization on the urban heat island effects along the Washington–Baltimore corridor. *Journal of Applied Meteorology and Climatology*, 50(10), 2012-2029.

Zhang, G. J., Cai, M., & Hu, A. (2013). Energy consumption and the unexplained winter warming over northern Asia and North America. *Nature Climate Change*, 3(5), 466-470.

Zhang, Y., Zhang, J., Zhang, X., Zhou, D., & Gu, Z. (2021). Analyzing the Characteristics of UHI (Urban Heat Island) in Summer Daytime Based on Observations on 50 Sites in 11 LCZ (Local Climate Zone) Types in Xi'an, China. *Sustainability*, 13(1), 83.

Zhang, Y., Zhang, J., Zhang, X., Zhou, D., & Gu, Z. (2021). Analyzing the Characteristics of UHI (Urban Heat Island) in Summer Daytime Based on Observations on 50 Sites in 11 LCZ (Local Climate Zone) Types in Xi'an, China. *Sustainability*, 13(1), 83.

Zhao, L., Oppenheimer, M., Zhu, Q., Baldwin, J. W., Ebi, K. L., Bou-Zeid, E., ... & Liu, X. (2018). Interactions between urban heat islands and heat waves. *Environmental Research Letters*, 13(3), 034003.

Zhao, Z. C. (2011). Impacts of urbanization on climate change. 10,000 Scientific Difficult Problems: Earth Science (in Chinese), 10,000 scientific difficult problems Earth Science Committee Eds. Science Press, 843, 846.

Zhou, D., Zhao, S., Zhang, L., Sun, G., & Liu, Y. (2015). The footprint of urban heat island effect in China. *Scientific Reports*, 5(1), 1-11.

Ziaul, S., & Pal, S. (2020). Modeling the effects of green alternative on heat island mitigation of a meso level town, West Bengal, India. *Advances in Space Research*, 65(7), 1789-1802.

Chapter 3

The Relationship between Climate Change Impacts on Urban Heat Island and Water Demand for Urban Green Spaces

Raphael Chukwuka Nnachi[1,*], Anne W. M. Ng[2,3,†] and Muhammad Atiq Ur Rehman Tariq[4,5,‡]

[1]Faculty of Biological Sciences, Alex Ekwueme Federal University, Ndufu Alike Ikwo, Abakaliki, Nigeria
[2]College of Engineering, Information Technology and Environment, Charles Darwin University, Brinkin, Australia
[3]Energy and Resources Institute, Charles Darwin University, Brinkin, Australia
[4]College of Engineering and Science, Victoria University, Melbourne, Australia
[5]Institute for Sustainable Industries & Liveable Cities, Victoria University, Melbourne, Australia

Abstract

The high rate of urbanization has resulted in higher temperatures in many cities owing to the alteration in energy balance which has ultimately exacerbated climate change effects. Consequently, the ambient temperatures of many megacities in the world are continuously increasing, leading to severe heatwaves and increased Urban Heat Island (UHI). Several mitigatory measures have been considered, including increasing albedo, cool pavements, cool roofs, urban lawns, urban trees and green roofs. In particular, the use of vegetation and urban green

---

[*]Author's Email: nnachi.raphael@funai.edu.ng.
[†]Author's Email: anne.ng@cdu.edu.au.
[‡]Corresponding Author's Email: atiq.tariq@yahoo.com.

In: Urban Heat Islands Reexamined
Editors: Satya Prakash and Anne W. Ng
ISBN: 979-8-88697-215-3

space, particularly trees, is a more sustainable approach. However, meeting the water demand for irrigating urban green space is a key challenge, especially in arid and semi-arid regions. To maximize urban water usage, a balance between urban energy and water demand must be achieved. Hence, it is very vital to estimate the total water requirements for indoor and outdoor usage using water resource modelling tools. Within this context, this paper discusses the nexus between climate change effects, UHI and water demand for urban green spaces. This paper focuses on the various environmental and water modelling tools for estimating urban water demand for indoor and outdoor usage. Additionally, this paper provides sustainable water resource management strategies and approaches for mitigating UHI and climate change effects.

Keywords: urban heat island, climate change, urban green space, water demand, climate and water modelling framework

1. Introduction

Rapid urban development and climate changes during the last few decades have had notable effects on urban climate [1]. Consequently, impervious urban surfaces and a lack of trees result in warmer temperatures in the city due to changes in surface energy balance, thus resulting in the creation of the urban heat island (UHI) effect [2]. Climate change increases hot days by a similar amount for both urban and rural situations, including the frequency, intensity and duration of heatwaves [3]. This condition creates the need for planting more vegetation, particularly trees that embrace sustainable UHI mitigation and temperature control of surrounding local climate systems [4]. However, less attention to water allocation for trees and limited available water sources in semi-arid cities, such as Melbourne, have increased the pressure on the sustainability of those trees [5]. In this instance, regulatory agencies around the world actively promote Urban Forest Strategy, aimed to increase the sustainable tree canopy covers for reducing urban temperature [6]. However, this requires significant coordination between urban planning and water resource management using integrated urban water resources models (IUWRM) [7, 8]. These are cost-effective tools that enable coordination by integrating tree water demand and alternative water sources for appropriate technology selection.

Australia is one of the driest inhabited continents, characterised by extreme climatic variability and has the lowest average rainfall of any inhabited continent [9]. This is particularly evident in Victoria, which has seen a decline in autumn, winter and spring rainfalls over the past two decades and runoff into most of waterways is projected to decrease between 5% and 45% by 2030 and between 5% and 50% by 2070 [10, 11]. By 2070 drought frequency is likely to increase by between 10% and 80% in the southern half of Victoria [12, 13]. In response to these challenges, a growing body of academics and professionals are advocating alternative approaches centred on including local assets and service provision within existing urban energy balance, centralised urban water and wastewater management [14–17].

2. Urban Heat Islands (UHI) Issues: Effects and Mitigation

Rapid urbanisation is a major driver of environmental change because as the population increases, additional urban lands are required given that the existing ones are used more intensely for the provision of facilities and services [18]. This reallocation of land generates many problems for city dwellers as it involves the conversion or degradation of natural ecosystems, such as trees, which have immediate impacts on urban temperature [19]. Urban surfaces are covered by complex buildings, roads, and artificial materials, such as asphalt and concrete. These impervious urban materials lead to changes in the absorption and reflection of solar radiation, and thus the surface energy balance [2]. As a result, cities become relatively warmer places compared to the surrounding rural areas (Figure 1) [20]. This phenomenon is well known as the urban heat island (UHI) effect. For instance, on hot summer days, parts of urban Melbourne can be 4°C - 7°C degrees hotter than surrounding rural areas [21]. Despite this rate, UHI are more extreme in other countries, such as Japan. The largest UHI effect was notably recorded in Tokyo with an increase of 12°C [22]. Therefore, understanding the relationship between urban development and local climate is vital for urban planning and design to determine effective design strategies, such as altering the vegetation and irrigation regime to improve urban climate.

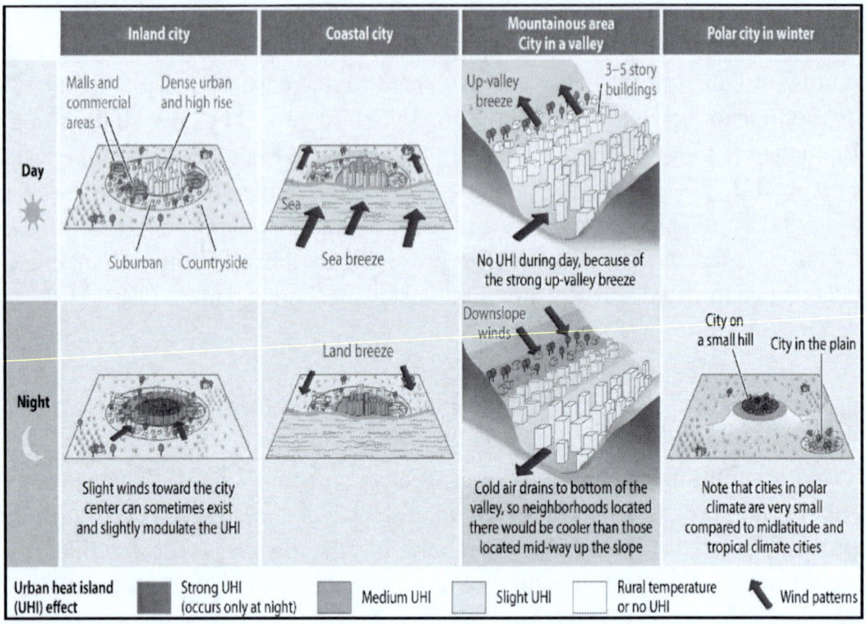

Figure 1. Schematics of the processes leading to the formation of Urban Heat Island under different climate and weather conditions [23].

2.1. Urban Heat Island Effects

2.1.1. High Mortality and Morbidity Rates

Several health related UHI consequences are beneficial, such as reduced winter mortality due to anthropogenic heat and less ice and snow on roads [24]. However, the majority of studies consistently found that human exposure to excessively warm weather, especially in cities, is an increasingly critical public health problem [25–28]. Heat-related problems are critical issue in many urban areas in the world [29]. Wong and his team reviewed the global UHIs and found that many of the reviewed areas have been linked to increased mortality risk as a result of the elevated urban temperatures during heatwaves [22]. For instance, an estimated 15,000 deaths were attributed to the heat wave event across France in August 2003 [30].

In 2009 and 2013, heatwaves in Victoria highlighted the risk of heat-stress related fatalities [31]. The 2009 heatwave caused 374 excess deaths and in 2013, there were approximately 400 heat related deaths in Victoria [32]. The elderly, young children, and people with chronic respiratory and cardiovascular diseases are physically more susceptible to the dangers of prolonged exposure to excessive heat than healthy young adults. With excessive increases in temperature, the associated increase in human body temperature exacerbates pre-existing health problems such as fever, heart diseases and severe heat stroke, potentially resulting in death in the worst cases. The impact of UHI is thus a serious issue in megacities cities such as Melbourne and therefore requires mitigation measures [33, 34].

2.1.2. High Energy Consumption

UHI results to a negative effect on energy consumption. It affects the energy performance of buildings located in densely built areas. In this case, higher temperatures increase energy use, specifically for air-conditioning usage in buildings [35]. Several studies have indicated that this has a significant effect on the energy consumption of buildings [36, 37] and it has been found that the peak urban electricity demand increases 2-4% for each 1K rise in daily maximum temperature [36]. Furthermore, about 5–10% of the current urban electricity demand is expended to cool buildings to compensate the 0.5°C – 3.0°C rise in urban temperatures. It is estimated that UHI increases power consumption by about 1.5GW, costing the ratepayers over $100 million per year [39]. In the United States (US) the additional air-conditioning usage caused by higher urban air temperature increase is responsible for 5–10% of urban peak electricity demand which costs several billion dollars annually. In Melbourne, Australia, UHI has an effect of increasing the cooling energy in Central Business Districts (CBD) by 46% and 23% for existing and newly constructed houses, respectively [40]. The lower rate for new houses is accounted for by better design, so any reduction in UHI rates would yield more economic benefits.

2.1.3. Poor Air Quality

The relationship between UHI and air quality has been investigated by several studies. Those studies demonstrate that a higher UHI intensity leads to higher air pollutant concentrations, such as O_3, and NO_2 [41–43]. In a recent study by Wang and his team to understand the effect of UHI and ozone pollution, they showed that UHI is a key factor affecting the O_3 level and they found that the average hourly O_3 concentration was 61.83 μg/m^3 (30.92 ppb) and the highest value was 105.32 μg/m^3 (52.66 ppb) at 15:00 pm (UHI effect) [44]. During summer period, the O_3 peak was 82.50 μg/m^3 (41.25 ppb). Understanding the relationship between UHI and air pollution is necessary for improving air quality. This is because rapid development of urbanization and industrialization processes has increased air pollution which has become serious issue in most cities [44, 45].

2.1.4. Increased Water Demand and Climate Change

Climate change refers to any long-term shifts or trends in climate over many decades. It has the capacity to modify the climatic potential for UHI, with the increase in the number of hot days (maximum temperature >35°C) by a similar amount for both urban and rural situations [46]. Such higher temperatures in cities due to UHI tend to result in an increased urban water demand. This increase is especially likely when temperatures are above 15°C. Most of the urban water demands in UHI affected urban areas, are for irrigation and other outdoor purposes. Several studies have investigated the relationship between UHI and increased water demand [47–49]. Each rise in temperature is equivalent to 1,318 litres, which is 3.8% of domestic water demand, and that 1°C reduction of diurnal temperature result in an increase of 1,722 litres in average residential water demand [47]. In terms of night-time temperature, Aggarwal and his team found that each degree of temperature increases around 1.4% in domestic water demand [50]. These findings confirmed that higher temperatures due to UHI effect increases the urban water demand for outdoor residential urban vegetation and recreation. In Australia, the impact of climate change has already occurred through rising temperatures with increasing frequency, intensity and duration of heatwaves, more intense bushfires, intense floodings and longer severe droughts [51]. Figure 2 illustrates the negative impact of climate change and its influence on the environment [52, 53]. In Melbourne, the summer temperature indicates that UHI effect can increase local suburban temperatures by several degrees [54]. Although this degree of increase depends on the surface albedo of the built environment, the amount of vegetation canopies and the possible existence of open water. According to CSIRO and BOM the temperature of Australian cities is projected to continue warming, rising by 0.6°C to 1.5°C by 2030 compared with the climate of 1980 to 1999 [55, 56] In addition, the warming level by 2070 compared to 1980-1999 is projected to be 1.0 to 2.5°C for low greenhouse gas emissions areas and 2.2 to 5.0°C for high emissions areas. Consequently, this can increase UHI intensity in major Australian cities in the future.

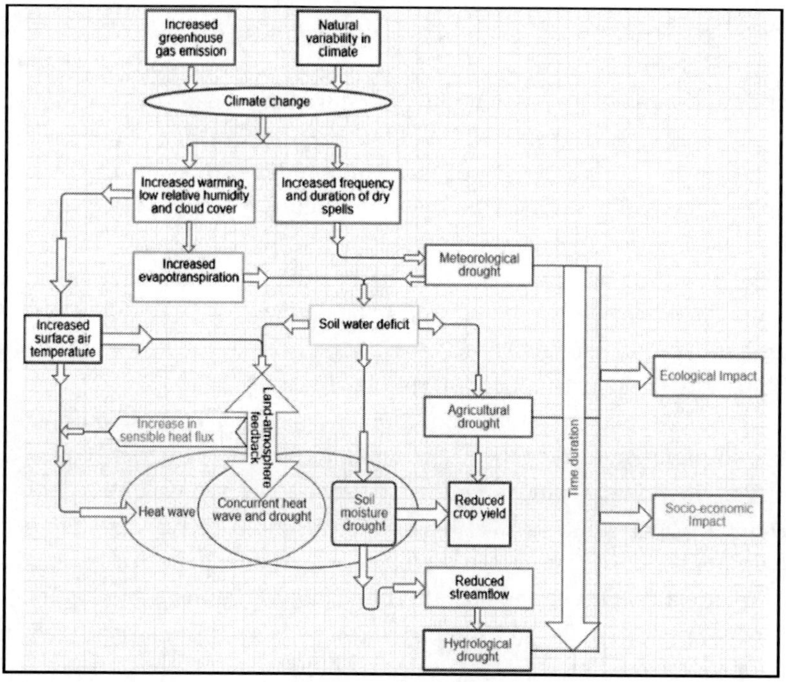

Figure 2. Impacts of climate change and water demand challenges in the environment [52, 53].

2.2. UHI Mitigation

Several UHI mitigation strategies have been investigated in the past. Since there are various UHI mitigation approaches available for modifying urban microclimate, it can be difficult to select the most suitable approach to implement. Characteristics of local climate and urban configurations play a significant role in the implementation of any selected approache [57]. Therefore, there is a need to gain in-depth understanding of each measure, particularly concerning its advantages and challenges in specific urban areas. Below are the commonest UHI mitigation approaches.

2.2.1. Open Water Bodies

Open water bodies have a potential to reduce UHI because they are naturally or artificially located in urban precincts. They include lakes, ponds, wetlands and rivers. The existence of these water bodies in the urban area lowers air temperature through evaporation, particularly during daytime [58]. In turn this reduces heat stress. Many studies have shown the microclimate benefits of open water bodies in reducing temperature and heat stress [59–61]. For example, during the warmest hours of the day, the presence of ponds drops the temperature by 1.6°C and reduced heat stress index by 0.8–1.1°C [62, 63]. [64] compared the reduction of air temperature between areas of land with a lake and areas without a lake and they found that the air temperature reduction near the lake could be as much as 3.5°C higher than the air temperature in area without a lake. According to the study, the cooling effect of a lake with 15°C water temperature can reach 1.5°C or 2°C in areas close to the lake and only 0.5°C or 1°C in the urban location further down the lake.

Although the above study found that open water bodies could generally reduce the temperature of their surroundings, this is particularly true during daytime. However, during nighttime when people are usually at home, the presence of open water bodies in urban area could increase air temperature and promote undesirable human thermal comfort due to the diurnal and annual cycle over water. During the late afternoon and nighttime, water temperatures become warmer than the surrounding surfaces. As a result, open water bodies warm their surroundings. [64] demonstrated that the presence of the lake leads to an increase in the air temperature (15°C) when water temperature was larger than surrounding temperature (20°C). These findings have consequences for using open water bodies in urban design, particularly when considering the additional microclimate effects.

2.2.2. Increasing Albedo

Albedo is defined as the proportion of the total incidence of light or radiation that is reflected by a surface back to space. Albedo values vary depending on urban surface types [65]. The typical average albedo values of urban surfaces range from 0.05 to 0.2 for asphalt roads, from 0.1 to 0.35 for concrete walls, from 0.2 to 0.4 for brick walls, from 0.1 to 0.35 for tile roofs [66, 67]. As such, increasing albedo of urban surfaces (i.e., roof, wall, and road) allows them to reflect a significant part of the incoming solar radiation or light. In terms of UHI mitigation studies, modification of those albedo values in urban space is quite strategic and now commonly implemented to two urban surfaces: roofs and pavements. Several studies have demonstrated the effectiveness of albedo increase to reduce the UHI at both daytime and night-time [57]. This is done by introducing cool materials which has highly reflective materials that can maintain lower surface temperatures and thus can present an effective solution to mitigate the UHI. However, there are some challenges to applying this strategy which can potentially lead to limited applications in urban areas as discussed below.

2.2.3. Cool Pavements

As mentioned previously, pavements are the major contributors to the development of UHI in urban area. During summertime, pavements tend to absorb and store solar radiation. To mitigate UHI effects caused by typical large use of pavements in urban precincts, the substitution of conventional pavement materials for cooler material is required [68]. Cool pavement materials increase albedo which assists in decreasing surface temperature as well as the amount of sensible heat released to the atmosphere. Recent studies and applications have shown that the cool pavements are closely associated with reductions in urban temperature [69]. For instance, Santamouris and his team found that the use of cool paving materials contributes to the reduction of the peak ambient temperature during a typical summer day in Greece, by 1.9°C [69]. In a study by Farhadi and his team, just substituting road materials from asphalt (with albedo 0.2 to 0.4) to concrete (0.2) can lower the heat capacity up to 35% [70]. In this case, more short-wave radiation is reflected, and less heat is stored in the ground surface of the cool pavement. Interestingly, they also found that the maximum ground surface temperature was reduced by 2°C in the detached area, 0.6°C in the middle-rise area, and 8.5°C in the high-rise area. At mid-night, cool pavement reduced urban temperature by 0.2°C.

However, while the use of high-albedo surfaces may lead to lower surface temperatures, more intense reflected radiation occurs during this process, which potentially contributes to pedestrian discomfort [71]. For example, Taleghani in his research demonstrated that the application of higher reflectivity of high-albedo materials in the open space of the courtyard increases average surface temperature during the day by 0.4°C [72]. The study concluded that an increase of every 0.1 albedo of the surfaces leads to 1.2°C higher mean radiant temperature, and consequently, 0.8°C higher potential evapotranspiration (PET). While cities develop and promote more sustainable urban infrastructures that encourage urban dwellers to walk or do more physical exercise for health reasons, such strategy could be controversial in implementation. This is because cool pavements can significantly affect pedestrians' thermal comfort in urban spaces [73].

2.2.4. Cool Roofs

Modification of roof surfaces has also been found to be an effective strategy to reduce UHI effects. This involves changing roof surface materials to both high solar reflectance and high thermal emittance. These materials increase the reflection of incoming solar radiation in urban areas by increasing the albedo of roof surfaces. For instance, when cool roofs are implemented, high albedo roofs can reach a reduction of 0.4°C-0.6°C in New York city. In addition, an increase in the roof albedo in the Baltimore-Washington metropolitan area from 0.3 to 0.7 can reduce the median roof surface temperature in the middle of the day by 9.6°C-11.3°C [74, 75]. Therefore, the Baltimore-Washington metropolitan area needs roofs that have an albedo value of 0.7 to reduce the surface UHI by 1°C, thereby suggesting the potentials of cool roofs to reduce the surface and near-surface UHIs. However, performances of cool roofs in temperature reduction are hindered by dirt accumulation on surface of cool roofs, leading to a higher cost of maintenance [76]. And, when compared to green roofs, cool roofs are still cheaper and easier to implement and install as a retrofit than green roofs. Hence, the cool roof strategy is a more viable and cost-effective approach for mitigating the city-scale UHI effect [76].

Others benefits of cool roofs have been widely discussed. These benefits include improved building comfort, energy and money savings, reduced roof maintenance, reduction of peak electricity demand and low levels of pollution [76]. [77] found that roof temperatures reach almost 70°C and about 50% of heat enters buildings through the roof slab in hot and arid climates. As a result, the use of cool roofs is closely associated to important reductions of the cooling load of the corresponding buildings. However, increasing the roof albedo has a major effect on the ambient temperature thereby reducing human thermal comfort [78]. The use of high-albedo materials for urban surfaces may reduce the air temperature to which pedestrians are exposed, which affects their radiant balance in surrounding environment. Although the application of such materials on building roofs is generally a 'win–win situation', the use of such materials in pavements or wall surfaces in pedestrian areas may be less desirable. Although this should not necessarily be construed as an objection to implementing high-albedo materials on a large scale in urban areas.

2.2.5. Urban Lawns

Urban lawn is part of urban spaces that are largely covered by grasses. Exposed grass has a significant cooling effect through evapotranspiration (ET) process. In this case, ET could use large amount of solar radiation as latent heat and reduce the air temperature and UHI intensity [79]. The cooling effect can be up to 2K in the case of vegetated surfaces, such as urban lawns [79]. Studies has shown that the ground surface temperature difference between lawns (with ET) and the nearby paved road (without ET) could be as much as 3°C in middle summer days [80]. However, they are considerably less effective in urban cooling when compared to trees. This difference in the two vegetation types is less obvious when the coverage ratio is less or equal to 16%. When the coverage ratio increases to 34% and above, planting trees at the ground level can reduce pedestrian-level air temperatures more notably than grass surfaces [81]. Although grass could be an effective strategy to mitigating UHI, the water consumption rate could pose a potential challenge. For instance, the water consumption rate by urban lawns is about 17.5–25 m^3/d and this gives it the lowest efficiency compared to trees or shrubs (with water demand rates of 12.5–25 m^3/d) [82]. Therefore, this strategy of UHI mitigation is not particularly suited to some Australian cities, particularly Melbourne as it requires too much water.

2.2.6. Urban Trees

Amongst the urban vegetation-based approaches, increasing tree canopy/cover is one of the most cost-effective ways to reduce urban heat islands. The microclimatic cooling effect of trees is obtained from evaporative cooling and, more importantly from direct shading to surfaces, which is more effective in reducing the radiant temperature [83]. This cooling effect of urban trees has been recorded in several studies, with reductions in air temperature of up to 3–4°C observed at mid-day during summer in the case of trees in streets and parks [79]. Recently, [84] measured air temperature in the daytime and found that tree cover could reduce air temperature at the tree level (4°C at 20 m above the ground), as well as higher level (2°C at 60 m above the ground). Similarly, Klemm and his team found that 10% tree cover in street lowers the mean radiant temperature by 1K [85].[83] compared a nearby shaded and unshaded open space and found that air temperatures under the canopies were 0.64°C-2.52°C lower. Although green roofs and urban lawns rely on evapotranspiration in reducing urban temperatures, planting trees at the ground level can reduce pedestrian-level air temperatures more notably than grass surfaces. It has a direct impact on levels of human comfort. In addition, planting trees has the direct effect of reducing atmospheric Carbon monoxide (CO) because each individual tree directly sequesters CO from the atmosphere through photosynthesis. This is significant as it reduces the probability of asthma and other respiratory conditions.

Furthermore, social, economic and environmental benefits of urban trees, are increasingly recognised in urban planning and policy [86]. Benefits include reduced energy consumption [87], improved air quality [88], carbon storage and sequestration [89], recreational opportunities, increased attractiveness of communities, improved wildlife habitat [90]. Due to the above reasons, more and more cities are actively promoting tree planting in their urban planning policy, such as Urban Forest Strategy. Furthermore, the use of trees in urban areas is a growing strategy for future greener cities worldwide, especially in semi-arid regions. Most cities so far have focused on planting trees in parks and streets. For example, City of Melbourne has been planning to plant 1 million trees in its urban green space areas by end of 2018 [91]. Intensive cultivation of trees in urban parks and streets is increasing, however, there are challenges in establishing and maintaining sustainable healthy trees, such as the quantity of water required. This is a key challenge, particularly in semi-arid cities and is yet to be explored. Healthy trees are highly dependent on an adequate amount of water, especially public trees in the street and park, which have been in the past always subject to water restrictions during drought. Therefore, a better understanding of reasonable volume of water needed and water planning strategy is essential for future management of urban trees in attempt to gain and maintain long term benefits to the urban dwellers.

2.2.7. Green Roof

Green roof is a roof of a building which is partially or completely covered with vegetation. This strategy can greatly reduce rooftop surface and ambient air températures by increasing the evapotranspiration through soil and plants on rooftops [38]. The effectiveness and benefits of green roofs for such purpose has been studied extensively. Research has found the benefits to include decreased energy consumption, better air quality, noise reduction, and increased durability of the roof materials [92]. For example, [38] simulated the application of green roof on a city scale and found that green roofs can reduce the average ambient temperature between 0.3 and 3°C. Similarly, [76] found that increasing green roof fractions can significantly reduce the daytime surface UHI at around 4°C, but only moderately reduces the nighttime surface UHI (≈1°C). Recent study conducted by Sharma and his team demonstrated that daytime roof temperature reduced and varied linearly with increasing green roof fractions, from less than 1°C for the case of 25% green roof to~3°C during peak daytime for 100% green roof [93]. However, a case study from city of Hongkong, which has a high building-height-to-street-width (H/W) ratio, revealed that roof greening is ineffective for human thermal comfort near the ground [81]. Also, green roofs display low biodiversity as many green roofs do not contribute much to diversity of plant species or animal habitat.

3. Frameworks for Urban Water Demand and Climate Modelling

3.1. Urban Water Modelling

To investigate the integration of water provision for urban green space, researchers are required to use modelling to estimate the volume of water required and its source [94]. The following models are the most described in the literature of urban water resources modelling. Figure 3 depicts a flowchart of various parameters and considerations for optimal tool selection and modelling, while Table 3 shows the summary of the various tools for modelling urban energy and water balance.

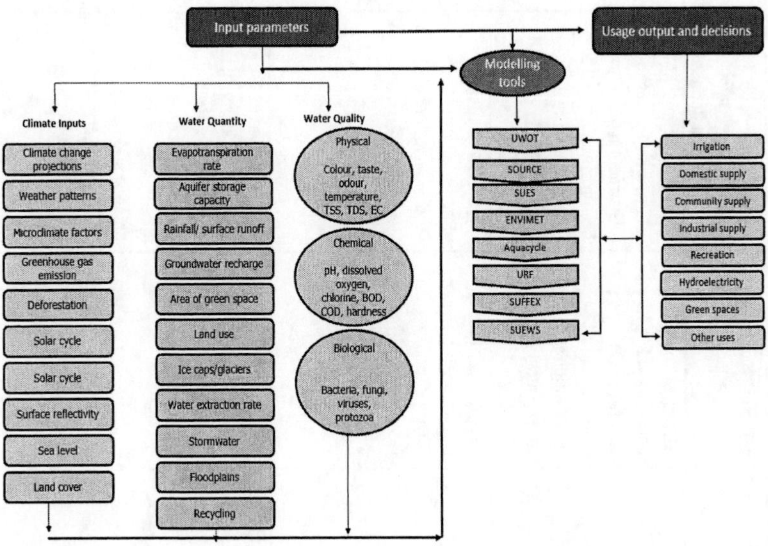

Figure 3. Water demand modelling tools for optimal water supply technology selection.

Table 1. Summary of analytical water and environmental modelling tools

Modelling tool	Principle of detection	Analytical scope/area of application in water resource management	Merits	Drawbacks	References
UWOT	Meeting urban water demands while ensuring safe management of wastewater and stormwater for proper planning	-Water recycling -Rainwater harvesting -Abstraction optimization -Flow-pattern restoration -Portable water demand reduction	-Estimates energy requirements for water appliances -Estimates stormwater and wastewater volumes -Assesses urban portable water demand alternatives	-May not account for all parameters such as the capital cost metrics, for capacity selection	[95, 96, 97, 98]
SUES	Estimation of urban evapotranspiration across various weather dynamics	-Run-off volumes -Energy requirements -External water balance -Rainfall and irrigation transformation	-Appropriate for modelling effects of urban vegetation on carbon sequestration and microclimate -Appropriate for linking urban water and energy balance	-Mainly for external urban water balance modelling -Uses calibration parameters optimized for suburban catchment	[99, 100]
AQUACULTURE	Balancing hydrology and urban water cycle (water demand, supply and recycling)	-Identification of the impact of water sensitive urban design dynamics on water balance -Prediction of water balance and energy usage on an array of urban design conditions	-Very comprehensive; can simulate internal and external features of urban water cycle -Utilizes indirect energy constraint	-Does not represent wetland vegetation, so wetland is modelled as open water body as such cannot give precise evaporation rate from water surface -Functions with a uniform water use data across catchments and the model does not take into cognizance wastewater inflow into stormwater drainage	[100, 101]

Table 1. (Continued)

Modelling tool	Principle of detection	Analytical scope/area of application in water resource management	Merits	Drawbacks	References
WRF	Using global weather information for local weather prediction	-Prediction of local weather conditions such as precipitation for flood forecasting -Studying planetary atmosphere for terrestrial forecasting -Global and regional climate research	-Various optional (between 1h to 12h) rapid update cycle procedure -Improves the rainfall quantity and its temporal variation in a storm event which in turn helps in real time flood forecasting	-Less accurate rainfall prediction for the first 3h – 6h of precipitation -The accuracy of weather events (such as rainfall) is negatively affected by spin up effect, initial and lateral boundary conditions of the model	[108, 109]
SURFEX	Limited-area numerical weather prediction using computed weather and climate data	-Studying urban heat island effects -Investigation of mesoscale climate change impacts and town energy balance in cities and urban areas	-Provides a more realistic and detailed UHI results -Can be applied in offline mode for atmospheric feedback study used in analysing Urban heat island evolution -Very comprehensive, user friendly and incorporates various soil and vegetation interactions thereby does not require a separate model for additional computations	-Has the tendency to generate a higher maximum temperature at high-elevation stations in time of winter daytime -Has the tendency to cause local concentration of rainfall thereby decreasing the total accumulated precipitation during summertime	[110, 111, 112]

Modelling tool	Principle of detection	Analytical scope/area of application in water resource management	Merits	Drawbacks	References
SOURCE	Capturing physical and regulatory process and management rules of a regulated catchment	-Conflict and dispute resolution in a multi-jurisdictional catchment area -Used by many actors for effective water resource management particularly when there are competing interests	-Provides a peculiar understanding of the prominent features of a catchment at source points -Aids policy makes and water actors in national and international decision making -Encourages local community input and management of key deliverables and priorities	-Requires robust knowledge about large catchment -Severe hydrological events such as droughts could put pressure on the model for quick results and new guidelines for water management in such extreme cases	[102, 103, 104]
ENVI-MET	Simulates urban climates in cities for proper environmental planning	-Urban microclimate simulation -Heat mitigation analysis.	-Helps in improving thermal comfort through reduction of air temperature and surface albedo -Easy to use, and gives information regarding vegetation and green space on microclimate up to resolution of 1 m	-Subject to socio-cultural filtration of policy and decision making -Requires numerous validation studies under a range of urban climate conditions before result validation	[105, 106, 107]

Modelling tool	Principle of detection	Analytical scope/area of application in water resource management	Merits	Drawbacks	References
SUEWS	Simulation of urban energy and water fluxes using commonly measured meteorological data about a surface cover	-Calculation of energy and water balance at neighbourhood scale -Estimation of anthropogenic heat flux in dry and hot seasons -Estimation of daily external water use such as irrigation and lawn watering -Effective in urban climate disaster mitigation and strategy assessment	-Simulates the urban surface energy balance together with the total urban hydrological cycle while considering irrigation and runoff processes -Can be run for several periods (as little as a day or less to many years) -Defines water profiles to ensure the appropriate evaluation of the impacts of diverse water restriction schedules	-Requires further validation when evaluating climate studies in tropical and sup-tropical urban areas -Prior knowledge of the development of an urban area is required for analysis	[113, 114, 115]

3.1.1. Urban Water Optioneering Tool (UWOT)

The most common model for urban water modelling traditionally used in many developed countries such as the United Kingdom (UK) has been the Urban Water Optioneering Tool (UWOT) [116]. It is a model that simulates the generation and routing of urban water demands to facilitate the planning and assessment of distributed interventions in the urban water cycle. The tool is based on a water balance model which allows the investigation of interactions between the major urban water cycle streams. UWOT integrates Simulink/MATLAB and Microsoft Excel (via Excel's Visual Basic for Applications (VBA) programming environment) into a computationally strong and user-friendly decision support tool [117]. A user-friendly interface allows users to configure the tool easily and to perform optimizations. The design of the user interface for UWOT considered issues of accessibility, flexibility, and interactiveness.

Two types of results are produced by the tool: (a) numerical values for all water streams, including potable water demand, wastewater and runoff generated, presented in total, daily and daily per capita values and plotted in bar diagrams to allow for a quick impression of the water balance of the simulated configuration; (b) a wider assessment created by obtaining values for the sustainability indicators for all the technologies present in the simulated configuration, for which the user has provided (context-specific) values for the particular situation by entering them into the technology library [117]. The tool has been successfully tested on a case study site in the UK. Unfortunately, this model has some limitations in regard to its applicability to integrate new nodes, such as vegetation water demand.

3.1.2. SUES

SUES is a comprehensive urban evapotranspiration-interception model developed by [99] in the UK. SUES focuses on the external water balance and the transformation of water inputs, such as rainfall and irrigation, into Evapotranspiration (E) and drainage. Representation of urban E in the model at a fine timescale enables the estimation of urban E under all conditions. The E model is based on the surface energy balance, using a Penman Monteith type model [118]. Therefore, SUES explicitly includes the urban energy balance to constrain the water balance simulations, meaning that urban water use is coupled to urban microclimates and energy consumption. SUES is applicable at the cluster through to catchment spatial scales and is typically implemented using a 5 min time step aggregated to an hour to capture the fine-scale dynamics of rainfall, drainage and evaporation [100]. Required data inputs for the model require hourly meteorology (net radiation, heat storage, humidity, temperature, wind speed and rainfall), and land surface cover (paved, built, coniferous trees, deciduous trees, irrigated grass and un-irrigated grass).

3.1.3. Aquacycle

Aquacycle is a daily urban water balance model which has been developed to simulate the total urban water cycle as an integrated system [101]. Aquacycle focuses more on urban water balance aspects, estimating the volume of water demand and available stormwater and wastewater in different spatial scales. Aquacycle can simulate the urban water balance at three spatial scales: a single unit block, a cluster and a whole catchment. It assumes that the supply of water to a plant is a linear function of available water in the root zone [119]. The maximum daily evapotranspiration is represented by the potential rate. Three groups of input data are required by Aquacycle: indoor water usage, climate, and physical characteristics. It uses a daily time step and recognizes three types of land use: residential, road and public open space. In Aquacycle, the amount of urban ET from pervious areas, and of evaporation from impervious surfaces, is calculated separately by using semi empirical equations including Penman–Monteith [119]. The usefulness of modelling tools such as Aquacycle is also to include the role of urban ET on microclimate which is only relevant to this project.

3.1.4. Source

SOURCE software package was developed by eWater CRC, Australia. by encompassing the key functionalities of IQQM, REALM and MSM-Bigmod [120]. SOURCE is built on top of a Microsoft .NET-based technology stack, with TIME as a base modelling framework providing the base simulation system. SOURCE is designed to capture physical and regulatory processes and management rules of a regulated river system. These issues are addressed in the design of seven major components within the SOURCE simulation engine: i) Catchment runoff; ii) River system network; iii) Interactions between river and groundwater systems; iv) Water quality; v) River regulation and storages; vi) Demands (urban, irrigation and environmental) and vii) Complex River management rules [120].

In the urban setting, SOURCE would be able to address city scale simulation of centralised versus decentralised sources of water including wastewater recycling, stormwater harvesting, desalination plants, rainwater tanks and the effects of water restrictions on overall system performance. For urban context, SOURCE also allows users to incorporate user-specific forecasting techniques through expression editor and plug-in functionalities. The system is extensible through expression editor and plug-in functionalities which is suitable for the project as the researcher is able to integrate new parameter which is not available in the model itself. For example, the expression editor provides the capability to define a value using an expression based on the current or past state of another part of the system.

3.2. Urban Climate Modelling

The result of chosen water demand model needs to be represented in an urban climate model to have more understanding of the model to accommodate an urban context which needs to address water and the temperature particularly in severe arid context of many cities. The following models are the most used in urban climate modelling [121].

3.2.1. ENVI MET

The ENVI-met is a modelling tool, which has been widely used in the urban context, it was specifically developed by the Research Group Climatology at Ruhr-University Bochum in Germany [105]. ENVI-met is mainly used for modelling surface–plant–air interactions and specifically designed for investigating changes to landscape and the built environment in the urban environment. To model the interaction between microclimate and urban surfaces, such as walls, pavements, and vegetation, the model utilizes a computational fluid dynamics (CFD) approach, employing the non-hydrostatic incompressible Navier–Stokes equations for the wind field, the k-epsilon turbulence model and a combined advection–diffusion equation with the Alternating Directly Implicit (ADI) solution technique [122]. This aspect is particularly useful in an investigation about the possible reduction of urban temperature using urban greening. ENVI-met model have been successfully applied in the simulations of the micro-meteorological and human-biometeorological impacts of building and street design as well as urban greening [123]. Effect of trees on microclimate can be simulated in the model. In this case, Envi-met model uses the Leaf Area Density (LAD) for the characterization of plants, derived from LAI at 10 intervals along the height of the plant [124].

3.2.2. Weather Research and Forecasting (WRF)

The US Weather Research and Forecasting (WRF) Model [125] is an atmospheric model designed for both research and numerical weather prediction (NWP). WRF is written primarily in Fortran, can be built with several compilers, and runs predominately on platforms with UNIX-like operating systems. Recent studies address the challenges in modelling the urban irrigation in WRF Model. For example, [126] developed an improved SLUCM to incorporate detailed hydrological processes, including evaporation from vegetated and engineered surfaces, sub-facet heterogeneity for water transport, and multilayer green roof systems. [127] introduced urban irrigation, the oasis effect, and anthropogenic latent heat into the Noah/SLUCM system. [126] implemented physically based parametrizations of urban hydrological processes into the single layer urban canopy model in the WRF model. The new single-layer urban canopy model features the integration of, (1) anthropogenic latent heat, (2) urban irrigation (3) evaporation from paved surfaces, and (4) the urban oasis effect.

3.2.3. SUFFEX

SURFEX ("Surface externalisée") is a more recent land surface modeling platform developed by [128]. SURFEX has been designed with a modular structure, which means that new elements of the sciences concerned can be easily introduced as new models or options. SURFEX can be run in offline mode (0-D or 2-D runs) or in coupled mode (from mesoscale models to numerical weather prediction and climate models) [110–112]. An assimilation mode is included for numerical weather prediction and monitoring. It covers all surface types (natural surfaces, town, inland water and oceans) and can be directly coupled to an atmospheric model. SURFEX has been progressively implemented and validated in various atmospheric models used either for process studies or for operational numerical weather prediction and climate runs. For simulating urban surfaces, SURFEX uses the TEB model [112], which is a commonly used urban energy balance model. Traditionally, TEB has been used for neighborhood scale modelling applications, but the integrated vegetation (TEB-Veg) allows for better simulation of processes inside an urban canyon [129]. This includes the integration of the irrigation regime for representation of sprinklers, which add extra precipitation to the foliage and soil surfaces inside the canyon. There is also the possibility in TEB-Veg to represent drip irrigation systems, which only adds the supply of water to the ground.

3.2.4. SUEWS

The Surface Urban Energy and Water Balance Scheme (SUEWS) is a relatively simple model that can simulate both energy and water fluxes and was developed in the UK [130]. The model is centered on the urban energy balance and urban water balance. SUEWS is applicable at the neighbourhood to city scale. To simulate urban water and energy balances, SUEWS requires common measured meteorological (solar radiation, air temperature, relative humidity, surface air pressure, wind speed and precipitation) data and information about the surface cover. It has inbuilt formulas for a combination of simulations [113]. In addition to being software that is able to address simulations, SUEWS also considers seven surface types: paved surfaces (including roads, pavements, car parks), buildings, evergreen trees and shrubs, deciduous trees and shrubs, grass, bare soil and open water (e.g., rivers, lakes, swimming pools, fountains). SUEWS uses several sub-models which are designed to minimize the number of input variables required [113]. The model contains several sub-models, for example, for net all-wave radiation, storage heat fluxes, anthropogenic heat fluxes and external irrigation.

3.3. Water Demand Estimation for Urban Green Space

To reduce UHI using urban green space (park and street trees), an estimation of the water demand is required to ensure that these trees remain healthy and green [131]. Generally, the demand for water largely comes from three (3) major sectors, namely, agriculture, industry and urban [132]. Among those sectors, agriculture and industry accounts for majority of the water demands. In total, the combined amount of water needed far outweighs the demand of water for urban use. Several studies have revealed that there is a significant proportion of water demand from plants, mostly grasses in urban residential areas. For example, urban vegetation in cities in the western part of the United States of America (USA) can consume up to 50% of the municipal water budget [133]. Additionally, and in response to UHI effects, many USA city planners now prefer to increase the number of trees in their urban planning policy, such as urban forest strategy to increase tree canopy covers. Melbourne, Australia has been advocating the planting of 1 million trees in its urban green space areas since 2013 [91]. However, very limited studies reveal the allocation of the volume of water required for this tree planting exercise. Therefore, in terms of UHI mitigation and water conservation, accurate estimation of water demand at city scale is critical to the design and management of sustainable cities, incorporating UHI issue. A better understanding of the water demand of green spaces is required to efficiently allocate water resources between the different green spaces.

Since early 2010s, the study on the water demand for urban green space has attracted a great attention and various researchers have tried to estimate the water use of urban trees, focusing primarily on those in the private realm, such as residential landscapes [134]. Water demand varies from day to day and from season to season, but few authors have estimated water use in public open spaces, such as parks [135] and street trees [136]. Table 2 shows the daily urban water demand for some trees across various seasons of the year [137]. The water use of maple trees could range from 8 L/day to 25 L/day [138]. This demonstrates that climate has a significant impact on the water use as that the trees in hotter cities require much more water. Compared to residential trees where volume of water needed can be supplied from outdoor metered irrigated water [139].

Table 2. Daily water requirements for urban trees (m^3/feddan/day) (Adapted from [137]

Periods	m^3/feddan/day				
	Apricot & pear	Apple	Citrus	Mango	Grapes
January	NA	NA	7.5	9.0	NA
February	NA	NA	9.3	11.2	NA
March	9.5	10.4	12.4	15.3	4.2
April	15.7	16.8	16	20.6	8.6
May	23.1	24.5	18.6	25.0	16.7
June	26.9	28.3	19.4	26.9	20.9
July	26.0	27.4	18.7	26.0	20.2
August	23.7	24.9	17.1	23.7	17.1
September	19.9	21.0	14.5	20.0	12.5
October	14.5	15.9	11.6	15.9	8.3
November	8.9	9.6	8.8	11.6	5.0
December	NA	NA	7.3	9.6	NA

3.4. Water Sources for Maintenance of Urban Green Space

The source of water for meeting the water required by the increasing number of urban trees is extremely crucial in terms of the sustainability of trees [140]. Generally, water is a limited resource in many parts of the world. This is particularly the case in cities located in arid and semi-arid regions, such as

Melbourne, Australia. Therefore, developing urban water resources in such dry areas has primarily focused on dealing with the issue of water security [141]. This problem is attributable to increased population and economic development. The main source of water for those purposes predominantly relies on two important traditional water sources including surface water (rivers, lakes, wetlands or artificial dams) and groundwater from underground aquifer in the form of wells and boreholes [142].

Surface water emanates from rainfall-runoff from the surrounding catchment area. As it rains, the run-off flows into rivers and onto the reservoirs. While traditional water sources are increasingly under pressures, supplemental water sources are required and play a vital role. The phenomena of UHI effects during heat wave event highlighted the importance of water in providing a liveable, cooler, greener city [143]. Using water for outdoor irrigation of green spaces allows the population to sustain a 'greener' landscape and mitigate the effects of the UHI. Only a few studies have explored the water use of urban trees [144]. If heat mitigation is a primary goal, then irrigation may be necessary to support vegetation health.

3.4.1. Groundwater for Irrigation

In contrast, groundwater is located below the earth's surface. This is where water from rain and rivers migrates through the ground and is stored in porous soils and rocks. To access it, groundwater is often withdrawn by constructing and operating extraction wells. Compared to surface water, groundwater is often cheaper, more convenient and less vulnerable to pollution, but its reliability is also dependent on rainfall as primary input [145]. However, due to climate change, relying on traditional water sources for future water supply is not a sustainable option. In addition, human activities contribute to water pollution, thereby reducing the availability of fresh water sources. Furthermore, more pressures on water availability will be commonplace as sustainable greener cities are promoted with massive planting of trees in the urban green spaces. Consequently, the provision of additional water sources is required for irrigating those trees [146].

Unarguably, maintaining healthy tree canopy covers generates social, economic, and environmental benefits, including UHI mitigation. The City of Melbourne is aiming to increase tree canopy cover in public streets and parks from 22% in 2014 to 40% by 2040, to help reduce UHI impacts on human health [147]. However, all these require water as a major component and as such, provision of a sufficient amount of water is necessary. Compared to grass and horticultural plants, trees in urban spaces are rarely irrigated, particularly in public spaces, such as parks and streets [148]. Park and street trees mostly rely on rainfall, soil moisture, and groundwater, with groundwater being the most vital. Bijoor and his team found that some irrigated trees can also utilize groundwater through their roots, and hence do not rely solely on irrigation water [149]. As a result, such trees might be able to withstand drought and/or water conservation measures. In contrast, they observed that un-irrigated urban trees had difficulties extracting adequate water from groundwater, leading to the development of very shallow root systems [150]. Such trees are then more susceptible to drought, leading to eventual death.

In arid regions which are not naturally forested, urban trees must be irrigated, and this allocation of water may come at the expense of other uses. Trees that are planted in private residential or commercial realm tend to survive longer and are greener because they benefit from outdoor irrigation where soil moisture remains high during irrigation [151]. In public urban green spaces such as street and parks, trees are much more vulnerable to drought and are subject to water restriction. Most cities have significantly cut irrigation to previously irrigated green space [152]. For example, in the greater Adelaide metropolitan area, a 60% reduction of previously irrigated green spaces has been observed since late 2000s [153].

3.4.2. Potable Water for Irrigation

Potable water, also known as drinking water, is water that is safe to drink or to use for food preparation. In general, raw untreated water is collected from a surface water source (such as an intake on a lake or a river) or from a groundwater source (such as a water well drawing from an underground aquifer) within the watershed that provides the water resource [154]. The raw water is transferred to the water purification facilities using uncovered aqueducts, covered tunnels or underground water pipes. The source, delivered to the point of consumption, is called potable water if it meets the water quality standards required for human consumption [155]. In general, the priority for potable water is to meet the drinking and health needs of citizens. For that reason, urban water supply accounts for the greatest consumption of potable water. For example, in Melbourne, this type of water use comprises 64% of Melbourne's total water use in 2016-17 [156], while non-residential and unmetered account of 25% and 11%, respectively [156].

Several studies found that significant volume of potable water for urban use/residential supply is used to irrigate outdoor residential gardens at the back and front yards of properties [157, 158]. For example, urban vegetation in residential areas in cities in the western U.S. consume up to 50% of the municipal water budget [159]. Approximately 14–30% of municipal water consumption is used for outdoors irrigation in California [160]. In Phoenix, 45% of the city's municipal total water deliveries are for outdoor usage, i.e., irrigation to support landscaping [144]. A similar percentage of water is also used in Australia where about 34% of total domestic water was used for gardens [161]. This is the second-highest use of water for gardens in the world after the USA and is perhaps attributed to the dry climate. Household landscape irrigation accounts for a large percentage of the water usage in many cities. For instance, landscaping irrigation accounts for 50% of total household water usage in semiarid regions [162]. Meanwhile, for general open space purposes, only about 8%-9% of the available potable water is used in Australian cities as diverse as Melbourne and Perth [163]. This includes both public and private (back and front gardens) open space, and even less water is allocated to trees in urban areas.

3.4.3. Alternative Water Sources for Irrigation

As potable water driven by traditional water sources becomes scarce and more expensive, diversifying water sources is increasingly required [164]. This is particularly the case in arid and semi-arid cities where water restriction is applied during drought. Climate change has resulted in a great deal of uncertainty in the future reliability of traditional water sources [141]. Expansion of tree canopy cover is a major management goal in cities to sustain a healthy greener landscape and mitigate the effects of the UHI. However, this requires fixed supplemental water to remain healthy and green and a wide range of viable alternative water source options has been described [165]. These includes household scale options, such as rainwater harvesting [166], city scale options, such as stormwater harvesting [167] and recycled water [168]. Figure 4 shows the household and city-scale water resource management options.

Many countries across the world-wide have been focussing on investing heavily in technological treatments to create alternative water sources to meet the demand for potable water. Previous studies on the use of those alternative water sources have been well documented [169–171]. However, most of the studies have focused on augmenting water supply for increased urban residential area. Since Australia is an arid country, there are only limited opportunities to harvest and store more catchment water, and to do so is likely to have significant environmental as well as capital costs [156]. For example, Mitchell and his team highlighted the use of stormwater and wastewater as a potential substitute for a portion of the fresh water from reticulated supply system [172]. They also presented a water balance model representing water flows through the urban water supply, wastewater and stormwater systems. In addition, water supply industries are starting to supply alternative water from various sources, including recycled water from treatment plants, and stormwater harvested from stormwater drains [173]. For example, City West Water, water supply industry in Melbourne, currently supplies approximately 2.5 GL of alternative water per annum to a range of customers across several schemes [174].

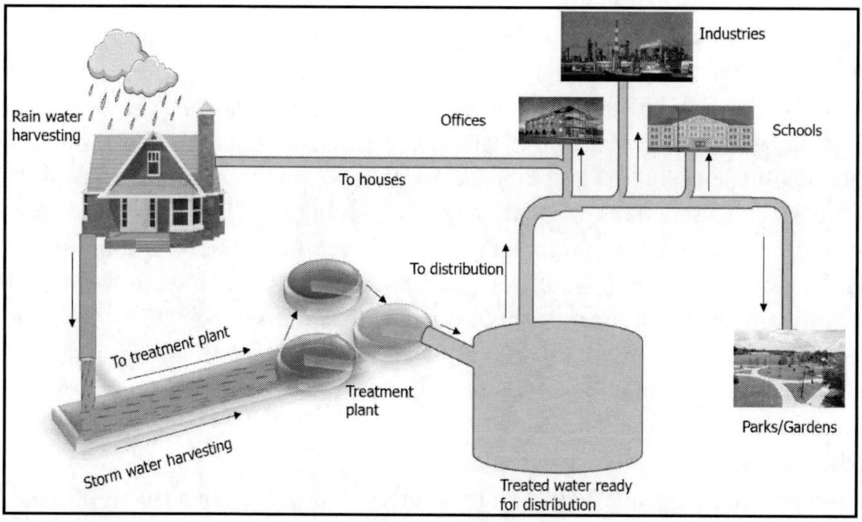

Figure 4. Rainwater and stormwater harvesting for urban water supply.

With the advancements in technology and increased requirement of alternative water supplies, a range of water sources such as stormwater, rainwater, and recycled water, have been introduced and are used in many parts of the world. Depending on the purpose, water is treated using appropriate technology and supplied to meet different water needs. Therefore, sustainability of alternative water sources in meeting these diverse water demands needs to be studied and understood in detail.

3.4.4. Rainwater Harvesting for Irrigation

Rainwater harvesting (RWH) is becoming a popular approach for alternative water resource at household scale. Persistent drought and the implementation of stringent water restrictions in semi-arid cities, such as Melbourne, have encouraged the use of rainwater harvesting within the property [175].

Rainwater harvesting refers to installation of rainwater tank to collect rainwater from roofs. The harvested rainwater is relatively less polluted than harvested stormwater [176] because it only passes over the roof surface before entering rainwater tanks. Therefore, it can be directly used for a range of non-potable uses, such as garden/lawn watering, cars washing, or toilet flushing, without any treatments. The use of this resource for non-potable purposes not only reduce the demand for potable water but also reduce the influent flows to the wastewater treatment plants, help mitigate sewer overflows and ultimately reduce the stress on the water and wastewater infrastructure [177].

In addition, harvested water is a low-cost alternative during dry periods, when potable water cost is higher, or water supply systems face severe shortages [178]. Furthermore, Gato-Trinidad and Gan (2014) found that implementation of rainwater tanks contributed to significant reduction (42.5%) in household water consumption [179]. In a consideration of RWH for park and street trees, several factors determine RWHS's efficiency. These include roof top areas, non-potable demand, rainfall patterns and the storage tank design [180]. Rainwater tank size and household size were found to be significant factors in the amount of household water savings [179]. Operation of rainwater tanks involves decentralised management reliant on individual tank owners operating them in appropriate ways to meet local flow targets.

3.4.5. Storm Water Harvesting

Another promising alternative water source for urban green space is harvested stormwater. Pervious surfaces provide a surface, allowing rainwater to infiltrate through the surface and into underlying layers, such as grass, planted areas, mulched groundcover, and porous pavements [181]. Traditionally, stormwater harvesting is simply collecting the rainwater run-off without any engineered treatment, releasing it into nearby rivers and creeks and then carrying the water into our bays. For certain cities, raw stormwater might be an abundant source of water in urban areas . In this case, stormwater runoff can be similar to, or even higher than the total potable water use. Stormwater runoff from Australian capital cities has been found to be comparable to the amount of potable water consumed [183]. For example, in 2010, Melbourne's annual potable water consumption sourced from water supply catchments was 356 GL, while the total available stormwater runoff was 463 GL [157]. The quantity of stormwater that can be harvested is highly dependent on the depth and intensity of rainfall. In addition, the capacity of the source area to shed or retain water also determines the volume of water that can be collected.

Moreover, the source area of harvested stormwater determines the quality of the stormwater supply and possible levels of treatment. As precipitation accumulates and flows over surfaces it collects pollutants and microbial contaminants, so to improve the quality, more advanced stormwater harvesting system is applied and refers to a constructed system that involves the capture, treatment, storage and use of stormwater runoff for beneficial use at a different time or place [184]. With the concept of water sensitive urban design (WSUD), a stormwater harvesting system is designable to meet water quantity and quality targets necessary for ecological health by adjusting the size of detention storages and outlets. WSUD involves technologies and approaches that aim to retain water in the urban landscape through stormwater harvesting and fit-for- purpose reuse and infiltration into soils to meet ecological, social and financial objectives. Currently, such decentralised approaches are commonplace throughout the world in limited applications for mitigating UHI [185]. Stormwater is being utilised by local governments, in arid and semi-arid areas, for parks and gardens, about 25% of the water used in Melbourne for parks and gardens is supplied by harvested stormwater [186]. Most regions in south Florida harvests stormwater during the wet season and employ it for irrigation and other uses during the dry period. Recently, storm water has been designed to supply sufficient volumes of water to urban trees through periods of drought via synthetic wicks from an underground storage tank to support adequate tree health [187].

3.4.6. Recycled Water for Irrigation

Recycled water has great potential to be water source for irrigating urban trees. The use of recycled water has grown worldwide over the past decade [188]. This is because recycled water has great opportunities to save valuable drinking water in an uncertain climate. Recycled water is derived from treated wastewater in wastewater treatment plants (WTP). With a certain degree of treatment, the water can be applied for a wide range of non-potable uses, including agriculture, industry, residential, and environment, such as contributing to river flows and recharging into aquifers [189]. Moreover, with the current advanced technology in WTP, it is possible to even generate drinking water. Compared to some other sources, there are benefits of using recycled water, such as consistency of supply regardless of season, climatic conditions and associated water restrictions. In some countries, it was found that treated wastewater is extensively used for irrigation. For instance, in Ghana [190], India [191], Pakistan [192], Vietnam [193], Mexico [194], Jordan and China [195].

Conclusion

As human population continues to increase, the water demand for residential, industrial and recreational purposes continues to increase and effective water management depends heavily on information about the quality and quantity of water available in a place. On the other hand, the Urban Heat Island effects remains one of the major environmental challenges in urban areas. Various mitigation approaches ranging from increasing surface reflectance or albedo using cool roofs to green roofs, cool pavements, etc have been proposed. Although these have a long-term benefit, but the high cost associated with implementation makes them less attractive options compared to more conventional options such as using urban vegetation.

With increasing climate change and variations in weather events, the water cycle is greatly impacted, and availability of water for maintaining and sustaining urban vegetation remains a challenge, particularly in dry regions. Therefore, there is a need to redefine water resource management approaches such as the use of technological framework for modelling water usage particularly in water-stressed areas. Surface and groundwater abstraction which are both the previous means of meeting urban water demand is now being supplemented with other alternative means such as rainwater harvesting, storm water harvesting and artificial recharge. This is very beneficial to the success of vegetation-based UHI mitigation measures, especially regarding drought periods common in Australian cities such as Melbourne. Such potential continuous supplemental irrigation from recycled water would be required to sustain evapotranspiration and shading, generating the microclimate beneficial effects of urban green spaces. More innovative water resource management techniques are required to meet the water demand for indoor and outdoor usage in urban areas as well as promotion of the efficient use of water.

References

[1] Miller, J. D., and M. Hutchins, "The impacts of urbanisation and climate change on urban flooding and urban water quality: A review of the evidence concerning the United Kingdom," *J. Hydrol. Reg. Stud.*, vol. 12, pp. 345–362, Aug. 2017, doi: 10.1016/j.ejrh.2017.06.006.

[2] Vujovic, S., B. Haddad, H. Karaky, N. Sebaibi, and M. Boutouil, "Urban Heat Island: Causes, Consequences, and Mitigation Measures with Emphasis on Reflective and Permeable Pavements," *Civil Eng*, vol. 2, no. 2, pp. 459–484, Jun. 2021, doi: 10.3390/civileng2020026.

[3] Tuholske, C., Caylor, K., Funk, C., Verdin, A., Sweeney, S., Grace, K., Peterson, P., & Evans, T. "Global urban population exposure to extreme heat," *Proc. Natl. Acad. Sci.*, vol. 118, no. 41, Oct. 2021, doi: 10.1073/pnas.2024792118.

[4] Shareef, S., "The Influence of Greenery and Landscape Design on Solar Radiation and UHI Mitigation: A Case Study of a Boulevard in a Hot Climate," *World*, vol. 3, no. 2, pp. 175–205, Mar. 2022, doi: 10.3390/world3020010.

[5] Costelloe, J. F., E. Payne, I. E. Woodrow, E. C. Irvine, A. W. Western, and F. W. Leaney, "Water sources accessed by arid zone riparian trees in highly saline environments, Australia," *Oecologia*, vol. 156, no. 1, pp. 43–52, May 2008, doi: 10.1007/s00442-008-0975-4.

[6] Davies, H. J., K. J. Doick, M. D. Hudson, and K. Schreckenberg, "Challenges for tree officers to enhance the provision of regulating ecosystem services from urban forests," *Environ. Res.*, vol. 156, pp. 97–107, Jul. 2017, doi: 10.1016/j.envres.2017.03.020.

[7] Wagner, I., and P. Breil, "The role of ecohydrology in creating more resilient cities," *Ecohydrol. Hydrobiol.*, vol. 13, no. 2, pp. 113–134, 2013, doi: 10.1016/j.ecohyd.2013.06.002.

[8] Pallavi, S., S. R. Yashas, K. M. Anilkumar, B. Shahmoradi, and H. P. Shivaraju, "Comprehensive Understanding of Urban Water Supply Management: Towards Sustainable Water-socio-economic-health-environment Nexus," *Water Resour. Manag.*, vol. 35, no. 1, pp. 315–336, Jan. 2021, doi: 10.1007/s11269-020-02734-x.

[9] Broich, M., Huete, A., Tulbure, M. G., Ma, X., Xin, Q., Paget, M., Restrepo-Coupe, N., Davies, K., Devadas, R., & Held, A. "Land surface phenological response to decadal climate variability across Australia using satellite remote sensing," *Biogeosciences*, vol. 11, no. 18, pp. 5181–5198, Sep. 2014, doi: 10.5194/bg-11-5181-2014.

[10] Cai, W., and T. Cowan, "Dynamics of late autumn rainfall reduction over southeastern Australia," *Geophys. Res. Lett.*, vol. 35, no. 9, p. L09708, May 2008, doi: 10.1029/2008GL033727.

[11] Olaka, L. A., J. O. Ogutu, M. Y. Said, and C. Oludhe, "Projected Climatic and Hydrologic Changes to Lake Victoria Basin Rivers under Three RCP Emission Scenarios for 2015–2100 and Impacts on the Water Sector," *Water*, vol. 11, no. 7, p. 1449, Jul. 2019, doi: 10.3390/w11071449.

[12] State Government of Victoria, "Victoria: State of the Environment," *Victoria State Environ.*, p. 622, 2013, [Online]. Available: http://ebooks.cambridge.org/ref/id/CBO9781107415324A009.

[13] Jin, C., *Climate change impacts on wetlands in Victoria and implications for research and policy*, no. 199. 2009.

[14] Xue, X., Cashman, S., Gaglione, A., Mosley, J., Weiss, L., Ma, X. C., Cashdollar, J., & Garland, J. "Holistic analysis of urban water systems in the Greater Cincinnati region: (1) life cycle assessment and cost implications," *Water Res. X*, vol. 2, p. 100015, Feb. 2019, doi: 10.1016/j.wroa.2018.100015.

[15] Sharma, A., S. Burn, T. Gardner, and A. Gregory, "Role of decentralised systems in the transition of urban water systems," *Water Supply*, vol. 10, no. 4, pp. 577–583, Sep. 2010, doi: 10.2166/ws.2010.187.

[16] Schellenberg, T., V. Subramanian, G. Ganeshan, D. Tompkins, and R. Pradeep, "Wastewater Discharge Standards in the Evolving Context of Urban Sustainability–The Case of India," *Front. Environ. Sci.*, vol. 8, Apr. 2020, doi: 10.3389/fenvs.2020.00030.

[17] Reymond, P., R. Chandragiri, and L. Ulrich, "Governance Arrangements for the Scaling Up of Small-Scale Wastewater Treatment and Reuse Systems – Lessons From India," *Front. Environ. Sci.*, vol. 8, Jun. 2020, doi: 10.3389/fenvs. 2020.00072.

[18] Hegazy, I. R. and M. R. Kaloop, "Monitoring urban growth and land use change detection with GIS and remote sensing techniques in Daqahlia governorate Egypt," *Int. J. Sustain. Built Environ.*, vol. 4, no. 1, pp. 117–124, Jun. 2015, doi: 10.1016/j.ijsbe.2015.02.005.

[19] Seifollahi-Aghmiuni, S., Z. Kalantari, G. Egidi, L. Gaburova, and L. Salvati, "Urbanisation-driven land degradation and socioeconomic challenges in peri-urban areas: Insights from Southern Europe," *Ambio*, vol. 51, no. 6, pp. 1446–1458, Jun. 2022, doi: 10.1007/s13280-022-01701-7.

[20] Zeleňáková, M., P. Purcz, H. Hlavatá, and P. Blišťan, "Climate Change in Urban Versus Rural Areas," *Procedia Eng.*, vol. 119, pp. 1171–1180, 2015, doi: 10.1016/j.proeng.2015.08.968.

[21] Heaviside, C., H. Macintyre, and S. Vardoulakis, "The Urban Heat Island: Implications for Health in a Changing Environment," *Curr. Environ. Heal. Reports*, vol. 4, no. 3, pp. 296–305, Sep. 2017, doi: 10.1007/s40572-017-0150-3.

[22] Wong, K. V., A. Paddon, and A. Jimenez, "Review of World Urban Heat Islands: Many Linked to Increased Mortality," *J. Energy Resour. Technol.*, vol. 135, no. 2, Jun. 2013, doi: 10.1115/1.4023176.

[23] Masson, V., A. Lemonsu, J. Hidalgo, and J. Voogt, "Urban Climates and Climate Change," *Annu. Rev. Environ. Resour.*, vol. 45, no. 1, pp. 411–444, Oct. 2020, doi: 10.1146/annurev-environ-012320-083623.

[24] Akbari, H., and R. Levinson, "Evolution of Cool-Roof Standards in the US," *Adv. Build. Energy Res.*, vol. 2, no. 1, pp. 1–32, Jan. 2008, doi: 10.3763/aber.2008.0201.

[25] Ncongwane, K. P., J. O. Botai, V. Sivakumar, and C. M. Botai, "A Literature Review of the Impacts of Heat Stress on Human Health across Africa," *Sustainability*, vol. 13, no. 9, p. 5312, May 2021, doi: 10.3390/su13095312.

[26] Rohat, G., J. Flacke, A. Dosio, H. Dao, and M. Maarseveen, "Projections of Human Exposure to Dangerous Heat in African Cities Under Multiple Socioeconomic and Climate Scenarios," *Earth's Futur.*, vol. 7, no. 5, pp. 528–546, May 2019, doi: 10.1029/2018EF001020.

[27] Hajat, S., M. O'Connor, and T. Kosatsky, "Health effects of hot weather: from awareness of risk factors to effective health protection," *Lancet*, vol. 375, no. 9717, pp. 856–863, Mar. 2010, doi: 10.1016/S0140-6736(09)61711-6.

[28] Hsu, A., G. Sheriff, T. Chakraborty, and D. Manya, "Disproportionate exposure to urban heat island intensity across major US cities," *Nat. Commun.*, vol. 12, no. 1, p. 2721, Dec. 2021, doi: 10.1038/s41467-021-22799-5.

[29] Allen, Simon K., Vicente Barros, Ian Burton, Diarmid Campbell-Lendrum, Omar-Dario Cardona, Susan L. Cutter, Opha Pauline Dube, Kristie L. Ebi, Christoper Field, John W. Handmer, Padma N. Lal, Allan Lavell, Katherine J. Mach, Michael D. Mastrandrea, Gordon A. Mcbean, Reinhard Mechler, Tom Mitchell, Neville Nicholls, Karen L. O'Brien, Taikan Oki, Michael Oppenheimer, Mark Pelling, Gian-Kasper Plattner, Roger S. Pulwarty, Sonia I. Seneviratne, Thomas Stocker, Maarten K. van Aalst, Carolina Vera, Thomas J. Wilbanks. "Summary for Policymakers," in *Managing the Risks of Extreme Events and Disasters to Advance Climate Change Adaptation*, vol. 9781107025, C. B. Field, V. Barros, T. F. Stocker, and Q. Dahe, Eds. Cambridge: Cambridge University Press, 2012, pp. 3–22.

[30] Fouillet, A., Rey, G., Wagner, V., Laaidi, K., Empereur-Bissonnet, P., Le Tertre, A., Frayssinet, P., Bessemoulin, P., Laurent, F., De Crouy-Chanel, P., Jougla, E., & Hémon, D."Has the impact of heat waves on mortality changed in France since the European heat wave of summer 2003? A study of the 2006 heat wave," *Int. J. Epidemiol.*, vol. 37, no. 2, pp. 309–317, Apr. 2008, doi: 10.1093/ije/dym253.

[31] Coates, L., J. van Leeuwen, S. Browning, A. Gissing, J. Bratchell, and A. Avci, "Heatwave fatalities in Australia, 2001–2018: An analysis of coronial records," *Int. J. Disaster Risk Reduct.*, vol. 67, p. 102671, Jan. 2022, doi: 10.1016/j.ijdrr.2021.102671.

[32] Pham, T., C. Young, N. Woodford, D. Ranson, C. M. F. Young, and J. E. Ibrahim, "Difference in the characteristics of mortality reports during a heatwave period: retrospective analysis comparing deaths during a heatwave in January 2014 with the same period a year earlier," *BMJ Open*, vol. 9, no. 5, p. e026118, May 2019, doi: 10.1136/bmjopen-2018-026118.

[33] Imran, H. M., J. Kala, A. W. M. Ng, and S. Muthukumaran, "Effectiveness of vegetated patches as Green Infrastructure in mitigating Urban Heat Island effects during a heatwave event in the city of Melbourne," *Weather Clim. Extrem.*, vol. 25, p. 100217, Sep. 2019, doi: 10.1016/j.wace.2019.100217.

[34] Imran, H. M., J. Kala, A. W. M. Ng, and S. Muthukumaran, "Impacts of future urban expansion on urban heat island effects during heatwave events in the city of Melbourne in southeast Australia," *Q. J. R. Meteorol. Soc.*, vol. 145, no. 723, pp. 2586–2602, Jul. 2019, doi: 10.1002/qj.3580.

[35] Guan, L., "Energy use, indoor temperature and possible adaptation strategies for air-conditioned office buildings in face of global warming," *Build. Environ.*, vol. 55, pp. 8–19, Sep. 2012, doi: 10.1016/j.buildenv.2011.11.013.

[36] Akbari, H., M. Pomerantz, and H. Taha, "Cool surfaces and shade trees to reduce energy use and improve air quality in urban areas," *Sol. Energy*, vol. 70, no. 3, pp. 295–310, 2001, doi: 10.1016/S0038-092X(00)00089-X.

[37] Santamouris, M., "On the energy impact of urban heat island and global warming on buildings," *Energy Build.*, vol. 82, pp. 100–113, Oct. 2014, doi: 10.1016/j.enbuild.2014.07.022.

[38] Santamouris, M., "Cooling the cities – A review of reflective and green roof mitigation technologies to fight heat island and improve comfort in urban environments," *Sol. Energy*, vol. 103, pp. 682–703, May 2014, doi: 10.1016/j.solener.2012.07.003.

[39] Goswam, Y., *Advances in the Development of Cool Materials for the Built Environment*. 2012.

[40] Lin, F.-Y., R.-L. Hwang, and T.-P. Lin, "Impact of Residential Energy Consumption on Urban Heat Island Effect in Tainan," in *Proceedings of the 16th IBPSA Conference*, 2019, pp. 3347–3352, doi: 10.26868/25222708.2019.210470.

[41] Swamy, G., S. M. S. Nagendra, and U. Schlink, "Urban heat island (UHI) influence on secondary pollutant formation in a tropical humid environment," *J. Air Waste Manage. Assoc.*, vol. 67, no. 10, pp. 1080–1091, Oct. 2017, doi: 10.1080/10962247.2017.1325417.

[42] Wang, Y., Z. Guo, and J. Han, "The relationship between urban heat island and air pollutants and them with influencing factors in the Yangtze River Delta, China," *Ecol. Indic.*, vol. 129, p. 107976, Oct. 2021, doi: 10.1016/j.ecolind.2021.107976.

[43] Lai, L.-W. and W.-L. Cheng, "Air quality influenced by urban heat island coupled with synoptic weather patterns," *Sci. Total Environ.*, vol. 407, no. 8, pp. 2724–2733, Apr. 2009, doi: 10.1016/j.scitotenv.2008.12.002.

[44] Wang, Y., H. Du, Y. Xu, D. Lu, X. Wang, and Z. Guo, "Temporal and spatial variation relationship and influence factors on surface urban heat island and ozone pollution in the Yangtze River Delta, China," *Sci. Total Environ.*, vol. 631–632, pp. 921–933, Aug. 2018, doi: 10.1016/j.scitotenv.2018.03.050.

[45] Li, J., M. Georgescu, P. Hyde, A. Mahalov, and M. Moustaoui, "Regional-scale transport of air pollutants: impacts of Southern California emissions on Phoenix ground-level ozone concentrations," *Atmos. Chem. Phys.*, vol. 15, no. 16, pp. 9345–9360, Aug. 2015, doi: 10.5194/acp-15-9345-2015.

[46] Wiesner, S., Bechtel, B., Fischereit, J., Gruetzun, V., Hoffmann, P., Leitl, B., Rechid, D., Schlünzen, K., & Thomsen, S."Is It Possible to Distinguish Global and Regional Climate Change from Urban Land Cover Induced Signals? A Mid-Latitude City Example," *Urban Sci.*, vol. 2, no. 1, p. 12, Jan. 2018, doi: 10.3390/urbansci2010012.

[47] Guhathakurta, S., and P. Gober, "Residential Land Use, the Urban Heat Island, and Water Use in Phoenix: A Path Analysis," *J. Plan. Educ. Res.*, vol. 30, no. 1, pp. 40–51, Sep. 2010, doi: 10.1177/0739456X10374187.

[48] Zipper, S. C., J. Schatz, C. J. Kucharik, and S. P. Loheide, "Urban heat island-induced increases in evapotranspirative demand," *Geophys. Res. Lett.*, vol. 44, no. 2, pp. 873–881, Jan. 2017, doi: 10.1002/2016GL072190.

[49] Syafii, N. I., M. Ichinose, N. H. Wong, E. Kumakura, S. K. Jusuf, and K. Chigusa, "Experimental Study on the Influence of Urban Water Body on Thermal Environment at Outdoor Scale Model," *Procedia Eng.*, vol. 169, pp. 191–198, 2016, doi: 10.1016/j.proeng.2016.10.023.

[50] Misra, A. K., "Climate change and challenges of water and food security," *Int. J. Sustain. Built Environ.*, vol. 3, no. 1, pp. 153–165, Jun. 2014, doi: 10.1016/j.ijsbe.2014.04.006.

[51] Cianconi, P., S. Betrò, and L. Janiri, "The Impact of Climate Change on Mental Health: A Systematic Descriptive Review," *Front. Psychiatry*, vol. 11, Mar. 2020, doi: 10.3389/fpsyt.2020.00074.

[52] Mukherjee, S., A. Mishra, and K. E. Trenberth, "Climate Change and Drought: a Perspective on Drought Indices," *Curr. Clim. Chang. Reports*, vol. 4, no. 2, pp. 145–163, Jun. 2018, doi: 10.1007/s40641-018-0098-x.

[53] Donald, A. W., *Drought as a Natural Hazard: Concepts and Definitions*. London: Routledge: Routledge, 2000.

[54] Yenneti, K., Ding, L., Prasad, D., Ulpiani, G., Paolini, R., Haddad, S., & Santamouris, M. "Urban Overheating and Cooling Potential in Australia: An Evidence-Based Review," *Climate*, vol. 8, no. 11, p. 126, Nov. 2020, doi: 10.3390/cli8110126.

[55] Ahmed, T., D. H. Vu, K. M. Muttaqi, and A. P. Agalgaonkar, "Load forecasting under changing climatic conditions for the city of Sydney, Australia," *Energy*, vol. 142, pp. 911–919, Jan. 2018, doi: 10.1016/j.energy.2017.10.070.

[56] Keywood, M., M. Hibberd, and K. Emmerson, *Australia state of the environment 2016: atmosphere, independent report to the Australian Government Minister for the Environment and Energy*. Canberra: Australian Government Department of the Environment and Energ, 2017.

[57] Wang, Y., U. Berardi, and H. Akbari, "Comparing the effects of urban heat island mitigation strategies for Toronto, Canada," *Energy Build.*, vol. 114, pp. 2–19, Feb. 2016, doi: 10.1016/j.enbuild.2015.06.046.

[58] Jacobs, C., L. Klok, M. Bruse, J. Cortesão, S. Lenzholzer, and J. Kluck, "Are urban water bodies really cooling?," *Urban Clim.*, vol. 32, p. 100607, Jun. 2020, doi: 10.1016/j.uclim.2020.100607.

[59] Albdour, M. S. and B. Baranyai, "Water body effect on microclimate in summertime: A case study from Pécs," *Pollack Period.*, vol. 14, no. 3, pp. 131–140, Dec. 2019, doi: 10.1556/606.2019.14.3.13.

[60] Manteghi, G., H. Bin Limit, and D. Remaz, "Water Bodies an Urban Microclimate: A Review," *Mod. Appl. Sci.*, vol. 9, no. 6, Feb. 2015, doi: 10.5539/mas.v9n6p1.

[61] Steeneveld, G. J., S. Koopmans, B. G. Heusinkveld, and N. E. Theeuwes, "Refreshing the role of open water surfaces on mitigating the maximum urban heat island effect," *Landsc. Urban Plan.*, vol. 121, pp. 92–96, Jan. 2014, doi: 10.1016/j.landurbplan.2013.09.001.

[62] Fung, C. K. W., and C. Y. Jim, "Influence of blue infrastructure on lawn thermal microclimate in a subtropical green space," *Sustain. Cities Soc.*, vol. 52, p. 101858, Jan. 2020, doi: 10.1016/j.scs.2019.101858.

[63] Saaroni, H. and B. Ziv, "The impact of a small lake on heat stress in a Mediterranean urban park: the case of Tel Aviv, Israel," *Int. J. Biometeorol.*, vol. 47, no. 3, pp. 156–165, May 2003, doi: 10.1007/s00484-003-0161-7.

[64] Theeuwes, N. E., A. Solcerová, and G. J. Steeneveld, "Modeling the influence of open water surfaces on the summertime temperature and thermal comfort in the city," *J. Geophys. Res. Atmos.*, vol. 118, no. 16, pp. 8881–8896, Aug. 2013, doi: 10.1002/jgrd.50704.

[65] Siegert, M. J., "Evidence of the Growth and Decay of Ice Sheets on Glaciated Landscapes," in *Reference Module in Earth Systems and Environmental Sciences*, Elsevier, 2014.

[66] Santamouris, M., "Using cool pavements as a mitigation strategy to fight urban heat island—A review of the actual developments," *Renew. Sustain. Energy Rev.*, vol. 26, pp. 224–240, Oct. 2013, doi: 10.1016/j.rser.2013.05.047.

[67] Li, H., J. Harvey, and A. Kendall, "Field measurement of albedo for different land cover materials and effects on thermal performance," *Build. Environ.*, vol. 59, pp. 536–546, Jan. 2013, doi: 10.1016/j.buildenv.2012.10.014.

[68] Mohajerani, A., J. Bakaric, and T. Jeffrey-Bailey, "The urban heat island effect, its causes, and mitigation, with reference to the thermal properties of asphalt concrete," *J. Environ. Manage.*, vol. 197, pp. 522–538, Jul. 2017, doi: 10.1016/j.jenvman.2017.03.095.

[69] Santamouris, M., Gaitani, N., Spanou, A., Saliari, M., Giannopoulou, K., Vasilakopoulou, K., & Kardomateas, T. "Using cool paving materials to improve microclimate of urban areas – Design realization and results of the flisvos project," *Build. Environ.*, vol. 53, pp. 128–136, Jul. 2012, doi: 10.1016/j.buildenv. 2012.01.022.

[70] Farhadi, H., M. Faizi, and H. Sanaieian, "Mitigating the urban heat island in a residential area in Tehran: Investigating the role of vegetation, materials, and orientation of buildings," *Sustain. Cities Soc.*, vol. 46, p. 101448, Apr. 2019, doi: 10.1016/j.scs.2019.101448.

[71] Erell, E., D. Pearlmutter, D. Boneh, and P. B. Kutiel, "Effect of high-albedo materials on pedestrian heat stress in urban street canyons," *Urban Clim.*, vol. 10, pp. 367–386, Dec. 2014, doi: 10.1016/j.uclim.2013.10.005.

[72] Taleghani, M., "Outdoor thermal comfort by different heat mitigation strategies- A review," *Renew. Sustain. Energy Rev.*, vol. 81, pp. 2011–2018, Jan. 2018, doi: 10.1016/j.rser.2017.06.010.

[73] Taleghani, M. and U. Berardi, "The effect of pavement characteristics on pedestrians' thermal comfort in Toronto," *Urban Clim.*, vol. 24, pp. 449–459, Jun. 2018, doi: 10.1016/j.uclim.2017.05.007.

[74] Mohegh, A., P. Rosado, L. Jin, D. Millstein, R. Levinson, and G. Ban-Weiss, "Modeling the climate impacts of deploying solar reflective cool pavements in California cities," *J. Geophys. Res. Atmos.*, vol. 122, no. 13, pp. 6798–6817, Jul. 2017, doi: 10.1002/2017JD026845.

[75] Wang, L., M. Huang, and D. Li, "Where Are White Roofs More Effective in Cooling the Surface?," *Geophys. Res. Lett.*, vol. 47, no. 15, Aug. 2020, doi: 10.1029/2020GL087853.

[76] Li, D., E. Bou-Zeid, and M. Oppenheimer, "The effectiveness of cool and green roofs as urban heat island mitigation strategies," *Environ. Res. Lett.*, vol. 9, no. 5, p. 055002, May 2014, doi: 10.1088/1748-9326/9/5/055002.

[77] Vardoulakis, E., D. Karamanis, and G. Mihalakakou, “Heat island phenomenon and cool roofs mitigation strategies in a small city of elevated temperatures,” *Adv. Build. Energy Res.*, vol. 8, no. 1, pp. 55–62, Jan. 2014, doi: 10.1080/17512549.2014.890537.

[78] Baniassadi, A., D. J. Sailor, P. J. Crank, and G. A. Ban-Weiss, “Direct and indirect effects of high-albedo roofs on energy consumption and thermal comfort of residential buildings,” *Energy Build.*, vol. 178, pp. 71–83, Nov. 2018, doi: 10.1016/j.enbuild.2018.08.048.

[79] Shashua-Bar, L., D. Pearlmutter, and E. Erell, “The cooling efficiency of urban landscape strategies in a hot dry climate,” *Landsc. Urban Plan.*, vol. 92, no. 3–4, pp. 179–186, Sep. 2009, doi: 10.1016/j.landurbplan.2009.04.005.

[80] Qiu, G., S. Tan, Y. Wang, X. Yu, and C. Yan, “Characteristics of Evapotranspiration of Urban Lawns in a Sub-Tropical Megacity and Its Measurement by the ‘Three Temperature Model + Infrared Remote Sensing’ Method,” *Remote Sens.*, vol. 9, no. 5, p. 502, May 2017, doi: 10.3390/rs9050502.

[81] Ng, E., L. Chen, Y. Wang, and C. Yuan, “A study on the cooling effects of greening in a high-density city: An experience from Hong Kong,” *Build. Environ.*, vol. 47, pp. 256–271, Jan. 2012, doi: 10.1016/j.buildenv.2011.07.014.

[82] Bhat, N. R., K. Majda, and A. M. Sakib, “Selection of Crops for Sustainable Utilization of Land and Water Resources in Kuwait,” *World J. Agric. Sci.*, vol. 5, no. 2, pp. 201–206, 2009, [Online]. Available: https://www.researchgate.net/publication/238079442_Selection_of_Crops_for_Sustainable_Utilization_of_Land_and_Water_Resources_in_Kuwait.

[83] Lin, B.-S. and Y.-J. Lin, “Cooling Effect of Shade Trees with Different Characteristics in a Subtropical Urban Park,” *Hort Science*, vol. 45, no. 1, pp. 83–86, Jan. 2010, doi: 10.21273/HORTSCI.45.1.83.

[84] Wang, Y. and H. Akbari, “The effects of street tree planting on Urban Heat Island mitigation in Montreal,” *Sustain. Cities Soc.*, vol. 27, pp. 122–128, Nov. 2016, doi: 10.1016/j.scs.2016.04.013.

[85] Emmanuel, R. and A. Loconsole, “Green infrastructure as an adaptation approach to tackling urban overheating in the Glasgow Clyde Valley Region, UK,” *Landsc. Urban Plan.*, vol. 138, pp. 71–86, Jun. 2015, doi: 10.1016/j.landurbplan.2015.02.012.

[86] Morison, P. J. and R. R. Brown, “Understanding the nature of publics and local policy commitment to Water Sensitive Urban Design,” *Landsc. Urban Plan.*, vol. 99, no. 2, pp. 83–92, Feb. 2011, doi: 10.1016/j.landurbplan.2010.08.019.

[87] Song, X. P., P. Y. Tan, P. Edwards, and D. Richards, “The economic benefits and costs of trees in urban forest stewardship: A systematic review,” *Urban For. Urban Green.*, vol. 29, pp. 162–170, Jan. 2018, doi: 10.1016/j.ufug.2017.11.017.

[88] Hewitt, C. N., K. Ashworth, and A. R. MacKenzie, “Using green infrastructure to improve urban air quality (GI4AQ),” *Ambio*, vol. 49, no. 1, pp. 62–73, Jan. 2020, doi: 10.1007/s13280-019-01164-3.

[89] Shadman, S., Ahanaf Khalid, P., Hanafiah, M. M., Koyande, A. K., Islam, Md. A., Bhuiyan, S. A., Sin Woon, K., & Show, P.-L."The carbon sequestration potential of urban public parks of densely populated cities to improve environmental sustainability," *Sustain. Energy Technol. Assessments*, vol. 52, p. 102064, Aug. 2022, doi: 10.1016/j.seta.2022.102064.

[90] Wood, E. M. and S. Esaian, "The importance of street trees to urban avifauna," *Ecol. Appl.*, vol. 30, no. 7, Oct. 2020, doi: 10.1002/eap.2149.

[91] Furlong, C., K. Phelan, and J. Dodson, "The role of water utilities in urban greening: A case study of Melbourne, Australia," *Util. Policy*, vol. 53, pp. 25–31, Aug. 2018, doi: 10.1016/j.jup.2018.06.005.

[92] Van Renterghem, T., "Green Roofs for Acoustic Insulation and Noise Reduction," in *Nature Based Strategies for Urban and Building Sustainability*, Elsevier, 2018, pp. 167–179.

[93] Sharma, A., P. Conry, H. J. S. Fernando, A. F. Hamlet, J. J. Hellmann, and F. Chen, "Green and cool roofs to mitigate urban heat island effects in the Chicago metropolitan area: evaluation with a regional climate model," *Environ. Res. Lett.*, vol. 11, no. 6, p. 064004, Jun. 2016, doi: 10.1088/1748-9326/11/6/064004.

[94] Saleem, A., I. Mahmood, H. Sarjoughian, H. A. Nasir, and A. W. Malik, "A Water Evaluation and Planning-based framework for the long-term prediction of urban water demand and supply," *Simulation*, vol. 97, no. 5, pp. 323–345, May 2021, doi: 10.1177/0037549720984250.

[95] Evangelos, R. and M. Christos, "Managing the complete Urban Water Cycle: the Urban Water Optioneering Tool," in *SWITCH Conference: The Future of Urban Water: Solutions for Livable and Resilient Cities*, 2011, pp. 1–2, doi: 10.13140/RG.2.1.3238.5682.

[96] Makropoulos, C., "Thinking platforms for smarter urban water systems: fusing technical and socio-economic models and tools," *Geol. Soc. London, Spec. Publ.*, vol. 408, no. 1, pp. 201–219, 2017, doi: 10.1144/SP408.4.

[97] Rozos, E., C. Makropoulos, and Č. Maksimović, "Rethinking urban areas: an example of an integrated blue-green approach," *Water Supply*, vol. 13, no. 6, pp. 1534–1542, Nov. 2013, doi: 10.2166/ws.2013.140.

[98] Makropoulos, C., E. Rozos, and D. Butler, "Urban water modelling and the daily time step: issues for a realistic representation," *8th International Conference on Hydroinformatics 2009, Concepcion, Chile, Curran Associates, Inc., USA,* 2011.

[99] Grimmond, C. S. B. and T. R. Oke, "An evapotranspiration-interception model for urban areas," *Water Resour. Res.*, vol. 27, no. 7, pp. 1739–1755, Jul. 1991, doi: 10.1029/91WR00557.

[100] Mitchell, V. G., H. A. Cleugh, C. S. B. Grimmond, and J. Xu, "Linking urban water balance and energy balance models to analyse urban design options," *Hydrol. Process.*, vol. 22, no. 16, pp. 2891–2900, Jul. 2008, doi: 10.1002/hyp.6868.

[101] de Carvalho, J. W. L. T. and I. Dos Santos, "Urban Configuration and Water Balance with Aquacycle Model in the Belém Catchment, Southern Brazil," *Caminhos Geogr.*, vol. 22, no. 79, pp. 149–160, Feb. 2021, doi: 10.14393/RCG227953860.

[102] Yang, A., G. Podger, S. Seaton, and R. Power, "A river system modelling platform for Murray-Darling Basin, Australia," *J. Hydroinformatics*, vol. 15, no. 4, pp. 1109–1120, Oct. 2013, doi: 10.2166/hydro.2012.153.

[103] Borrelli, M., "Review of Bridges Over Water: Understanding Transboundary Water Conflict, Negotiation and Cooperation," *J. Polit. Sci. Educ.*, vol. 8, no. 1, pp. 102–103, Jan. 2012, doi: 10.1080/15512169.2012.641396.

[104] Wheeler, K. G., C. J. Robinson, and R. H. Bark, "Modelling to bridge many boundaries: the Colorado and Murray-Darling River basins," *Reg. Environ. Chang.*, vol. 18, no. 6, pp. 1607–1619, Aug. 2018, doi: 10.1007/s10113-018-1304-z.

[105] Sahar, S., L. Ines, and C. Ulrich, "Using the ENVI-MET program to simulate the micro climate in new Town HASHTGERD," in *The international conference on Computing, Networking and Digital Technologies*, 2012, pp. 61–64, doi: 10.13140/2.1.1739.2005.

[106] Ozkeresteci, I., K. Crewe, A. J. Brazel, and M. Bruse, "Use and evaluation of the ENVI-met model for environmental design and planning. An Experiment on Lienar Parks," *South Africa*, no. August, pp. 10–16, 2003.

[107] Ayyad, Y. N. and S. Sharples, "Envi-MET validation and sensitivity analysis using field measurements in a hot arid climate," *IOP Conf. Ser. Earth Environ. Sci.*, vol. 329, no. 1, p. 012040, Sep. 2019, doi: 10.1088/1755-1315/329/1/012040.

[108] Liu, J., M. Bray, and D. Han, "A study on WRF radar data assimilation for hydrological rainfall prediction," *Hydrol. Earth Syst. Sci.*, vol. 17, no. 8, pp. 3095–3110, Aug. 2013, doi: 10.5194/hess-17-3095-2013.

[109] Yáñez-Morroni, G., J. Gironás, M. Caneo, R. Delgado, and R. Garreaud, "Using the Weather Research and Forecasting (WRF) Model for Precipitation Forecasting in an Andean Region with Complex Topography," *Atmosphere (Basel).*, vol. 9, no. 8, p. 304, Aug. 2018, doi: 10.3390/atmos9080304.

[110] Hamdi, R., P. Termonia, and P. Baguis, "Effects of urbanization and climate change on surface runoff of the Brussels Capital Region: a case study using an urban soil-vegetation-atmosphere-transfer model," *Int. J. Climatol.*, vol. 31, no. 13, pp. 1959–1974, Nov. 2011, doi: 10.1002/joc.2207.

[111] Hamdi, R., Degrauwe, D., Duerinckx, A., Cedilnik, J., Costa, V., Dalkilic, T., Essaouini, K., Jerczynki, M., Kocaman, F., Kullmann, L., Mahfouf, J.-F., Meier, F., Sassi, M., Schneider, S., Váňa, F., & Termonia, P. "Evaluating the performance of SURFEXv5 as a new land surface scheme for the ALADINcy36 and ALARO-0 models," *Geosci. Model Dev.*, vol. 7, no. 1, pp. 23–39, Jan. 2014, doi: 10.5194/gmd-7-23-2014.

[112] Gabriella, Z. and S. Gabriella, "Investigating the urban climate characteristics of two Hungarian cities with SURFEX/TEB land surface model," in *9 th International Conference on Urban Climate jointly with 12th Symposium on the Urban Environment*, 2015, pp. 1–5.

[113] Järvi, L., C. S. B. Grimmond, and A. Christen, "The Surface Urban Energy and Water Balance Scheme (SUEWS): Evaluation in Los Angeles and Vancouver," *J. Hydrol.*, vol. 411, no. 3–4, pp. 219–237, Dec. 2011, doi: 10.1016/j.jhydrol.2011.10.001.

[114] Demuzere, M., Coutts, A. M., Göhler, M., Broadbent, A. M., Wouters, H., van Lipzig, N. P. M., & Gebert, L. "The implementation of biofiltration systems, rainwater tanks and urban irrigation in a single-layer urban canopy model," *Urban Clim.*, vol. 10, pp. 148–170, Dec. 2014, doi: 10.1016/j.uclim.2014.10.012.

[115] Ao, X., Grimmond, C. S. B., Ward, H. C., Gabey, A. M., Tan, J., Yang, X.-Q., Liu, D., Zhi, X., Liu, H., & Zhang, N. "Evaluation of the Surface Urban Energy and Water Balance Scheme (SUEWS) at a Dense Urban Site in Shanghai: Sensitivity to Anthropogenic Heat and Irrigation," *J. Hydrometeorol.*, vol. 19, no. 12, pp. 1983–2005, Dec. 2018, doi: 10.1175/JHM-D-18-0057.1.

[116] Peña-Guzmán, C. A., J. Melgarejo, D. Prats, A. Torres, and S. Martínez, "Urban Water Cycle Simulation/Management Models: A Review," *Water*, vol. 9, no. 4, p. 285, Apr. 2017, doi: 10.3390/w9040285.

[117] C. K. Makropoulos, K. Natsis, S. Liu, K. Mittas, and D. Butler, "Decision support for sustainable option selection in integrated urban water management," *Environ. Model. Softw.*, vol. 23, no. 12, pp. 1448–1460, Dec. 2008, doi: 10.1016/j.envsoft. 2008.04.010.

[118] "Two-Source Energy Balance Model to Calculate E, T, and ET: Comparison of Priestley-Taylor and Penman-Monteith Formulations and Two Time Scaling Methods," *Trans. ASABE*, pp. 479–498, May 2014, doi: 10.13031/trans.57.10423.

[119] Mitchell, V. G., R. G. Mein, and T. A. McMahon, "Modelling the urban water cycle," *Environ. Model. Softw.*, vol. 16, no. 7, pp. 615–629, Nov. 2001, doi: 10.1016/S1364-8152(01)00029-9.

[120] Welsh, W. D., Vaze, J., Dutta, D., Rassam, D., Rahman, J. M., Jolly, I. D., Wallbrink, P., Podger, G. M., Bethune, M., Hardy, M. J., Teng, J., & Lerat, J. "An integrated modelling framework for regulated river systems," *Environ. Model. Softw.*, vol. 39, pp. 81–102, Jan. 2013, doi: 10.1016/j.envsoft.2012.02.022.

[121] House-Peters, L. A. and H. Chang, "Urban water demand modeling: Review of concepts, methods, and organizing principles," *Water Resour. Res.*, vol. 47, no. 5, May 2011, doi: 10.1029/2010WR009624.

[122] Brozovsky, J., A. Simonsen, and N. Gaitani, "Validation of a CFD model for the evaluation of urban microclimate at high latitudes: A case study in Trondheim, Norway," *Build. Environ.*, vol. 205, p. 108175, Nov. 2021, doi: 10.1016/j.buildenv. 2021.108175.

[123] Crank, P. J., D. J. Sailor, G. Ban-Weiss, and M. Taleghani, "Evaluating the ENVI-met microscale model for suitability in analysis of targeted urban heat mitigation strategies," *Urban Clim.*, vol. 26, pp. 188–197, Dec. 2018, doi: 10.1016/j.uclim. 2018.09.002.

[124] Tsoka, S., "Investigating the Relationship Between Urban Spaces Morphology and Local Microclimate: A Study for Thessaloniki," *Procedia Environ. Sci.*, vol. 38, pp. 674–681, 2017, doi: 10.1016/j.proenv.2017.03.148.

[125] Powers, J. G., Klemp, J. B., Skamarock, W. C., Davis, C. A., Dudhia, J., Gill, D. O., Coen, J. L., Gochis, D. J., Ahmadov, R., Peckham, S. E., Grell, G. A., Michalakes, J., Trahan, S., Benjamin, S. G., Alexander, C. R., Dimego, G. J., Wang, W., Schwartz, C. S., Romine, G. S., … Duda, M. G. "The Weather Research and Forecasting Model: Overview, System Efforts, and Future Directions," *Bull. Am. Meteorol. Soc.*, vol. 98, no. 8, pp. 1717–1737, Aug. 2017, doi: 10.1175/BAMS-D-15-00308.1.

[126] Yang, J., Wang, Z.-H., Chen, F., Miao, S., Tewari, M., Voogt, J. A., & Myint, S."Enhancing Hydrologic Modelling in the Coupled Weather Research and Forecasting–Urban Modelling System," *Boundary-Layer Meteorol.*, vol. 155, no. 1, pp. 87–109, Apr. 2015, doi: 10.1007/s10546-014-9991-6.

[127] Miao, S. and F. Chen, "Enhanced modeling of latent heat flux from urban surfaces in the Noah/single-layer urban canopy coupled model," *Sci. China Earth Sci.*, vol. 57, no. 10, pp. 2408–2416, Oct. 2014, doi: 10.1007/s11430-014-4829-0.

[128] Albergel, C., Dutra, E., Munier, S., Calvet, J.-C., Munoz-Sabater, J., de Rosnay, P., & Balsamo, G. "ERA-5 and ERA-Interim driven ISBA land surface model simulations: which one performs better?," *Hydrol. Earth Syst. Sci.*, vol. 22, no. 6, pp. 3515–3532, Jun. 2018, doi: 10.5194/hess-22-3515-2018.

[129] Stavropulos-Laffaille, X., Chancibault, K., Andrieu, H., Lemonsu, A., Calmet, I., Keravec, P., & Masson, V. "Coupling detailed urban energy and water budgets with TEB-Hydro model: Towards an assessment tool for nature based solution performances," *Urban Clim.*, vol. 39, p. 100925, Sep. 2021, doi: 10.1016/j.uclim.2021.100925.

[130] Ward, H. C., S. Kotthaus, L. Järvi, and C. S. B. Grimmond, "Surface Urban Energy and Water Balance Scheme (SUEWS): Development and evaluation at two UK sites," *Urban Clim.*, vol. 18, pp. 1–32, Dec. 2016, doi: 10.1016/j.uclim.2016.05.001.

[131] Derkzen, M. L., A. J. A. van Teeffelen, and P. H. Verburg, "REVIEW: Quantifying urban ecosystem services based on high-resolution data of urban green space: an assessment for Rotterdam, the Netherlands," *J. Appl. Ecol.*, vol. 52, no. 4, pp. 1020–1032, Aug. 2015, doi: 10.1111/1365-2664.12469.

[132] Lv, H., Yang, L., Zhou, J., Zhang, X., Wu, W., Li, Y., & Jiang, D. "Water resource synergy management in response to climate change in China: From the perspective of urban metabolism," *Resour. Conserv. Recycl.*, vol. 163, p. 105095, Dec. 2020, doi: 10.1016/j.resconrec.2020.105095.

[133] Lowry, J. H., R. D. Ramsey, and R. K. Kjelgren, "Predicting urban forest growth and its impact on residential landscape water demand in a semiarid urban environment," *Urban For. Urban Green.*, vol. 10, no. 3, pp. 193–204, Jan. 2011, doi: 10.1016/j.ufug.2011.05.004.

[134] Grahn, P. and U. K. Stigsdotter, "The relation between perceived sensory dimensions of urban green space and stress restoration," *Landsc. Urban Plan.*, vol. 94, no. 3–4, pp. 264–275, Mar. 2010, doi: 10.1016/j.landurbplan.2009.10.012.

[135] Reyes-Paecke, S., J. Gironás, O. Melo, S. Vicuña, and J. Herrera, "Irrigation of green spaces and residential gardens in a Mediterranean metropolis: Gaps and opportunities for climate change adaptation," *Landsc. Urban Plan.*, vol. 182, pp. 34–43, Feb. 2019, doi: 10.1016/j.landurbplan.2018.10.006.

[136] Yi, K., Maxwell, J. T., Wenzel, M. K., Roman, D. T., Sauer, P. E., Phillips, R. P., & Novick, K. A."Linking variation in intrinsic water-use efficiency to isohydricity: a comparison at multiple spatiotemporal scales," *New Phytol.*, vol. 221, no. 1, pp. 195–208, Jan. 2019, doi: 10.1111/nph.15384.

[137] Samir, M. I., *Modern Irrigation Systems Operation and Maintenance*. Mariut: DSC-Mariut, Egypt, 2003.

[138] Wagner, Y., Pozner, E., Bar-On, P., Ramon, U., Raveh, E., Neuhaus, E., Cohen, S., Grünzweig, J., & Klein, T. "Rapid stomatal response in lemon saves trees and their fruit yields under summer desiccation, but fails under recurring droughts," *Agric. For. Meteorol.*, vol. 307, p. 108487, Sep. 2021, doi: 10.1016/j.agrformet.2021. 108487.

[139] Mostafavi, N., H. R. Shojaei, A. Beheshtian, and S. Hoque, "Residential Water Consumption Modeling in the Integrated Urban Metabolism Analysis Tool (IUMAT)," *Resour. Conserv. Recycl.*, vol. 131, pp. 64–74, Apr. 2018, doi: 10.1016/j.resconrec.2017.12.019.

[140] Falkenmark, M. and D. Molden, "Wake Up to Realities of River Basin Closure," *Int. J. Water Resour. Dev.*, vol. 24, no. 2, pp. 201–215, Jun. 2008, doi: 10.1080/07900620701723570.

[141] Herrera-Pantoja, M. and K. M. Hiscock, "Projected impacts of climate change on water availability indicators in a semi-arid region of central Mexico," *Environ. Sci. Policy*, vol. 54, pp. 81–89, Dec. 2015, doi: 10.1016/j.envsci.2015.06.020.

[142] Bouwer, H., "Integrated water management: emerging issues and challenges," *Agric. Water Manag.*, vol. 45, no. 3, pp. 217–228, Aug. 2000, doi: 10.1016/S0378-3774(00)00092-5.

[143] Ossola, A., G. D. Jenerette, A. McGrath, W. Chow, L. Hughes, and M. R. Leishman, "Small vegetated patches greatly reduce urban surface temperature during a summer heatwave in Adelaide, Australia," *Landsc. Urban Plan.*, vol. 209, p. 104046, May 2021, doi: 10.1016/j.landurbplan.2021.104046.

[144] Kaplan, S., S. W. Myint, C. Fan, and A. J. Brazel, "Quantifying Outdoor Water Consumption of Urban Land Use/Land Cover: Sensitivity to Drought," *Environ. Manage.*, vol. 53, no. 4, pp. 855–864, Apr. 2014, doi: 10.1007/s00267-014-0245-7.

[145] Thomas, C. W., J. W. Harvey, O. L. Franke, and W. M. Alley, *Groundwater and surface water: A single resource*, vol. 17, no. 5. Denver, Colorado: U.S. GEOLOGICAL SURVEY, 2005.

[146] Green, T. R., Taniguchi, M., Kooi, H., Gurdak, J. J., Allen, D. M., Hiscock, K. M., Treidel, H., & Aureli, A. "Beneath the surface of global change: Impacts of climate change on groundwater," *J. Hydrol.*, vol. 405, no. 3–4, pp. 532–560, Aug. 2011, doi: 10.1016/j.jhydrol.2011.05.002.

[147] Kaspar, J., D. Kendal, R. Sore, and S. J. Livesley, "Random point sampling to detect gain and loss in tree canopy cover in response to urban densification," *Urban For. Urban Green.*, vol. 24, pp. 26–34, May 2017, doi: 10.1016/j.ufug.2017.03.013.

[148] Pellegrini, P. and S. Baudry, "Streets as new places to bring together both humans and plants: examples from Paris and Montpellier (France)," *Soc. Cult. Geogr.*, vol. 15, no. 8, pp. 871–900, Nov. 2014, doi: 10.1080/14649365.2014.974067.

[149] Bijoor, N. S., H. R. McCarthy, D. Zhang, and D. E. Pataki, "Water sources of urban trees in the Los Angeles metropolitan area," *Urban Ecosyst.*, vol. 15, no. 1, pp. 195–214, Mar. 2012, doi: 10.1007/s11252-011-0196-1.

[150] Xylogiannis, E., A. Sofo, B. Dichio, G. Montanaro, and A. N. Mininni, "Root−to−Shoot Signaling and Leaf Water−Use Efficiency in Peach Trees under Localized Irrigation," *Agronomy*, vol. 10, no. 3, p. 437, Mar. 2020, doi: 10.3390/agronomy10030437.

[151] McCarthy, H. R. and D. E. Pataki, "Drivers of variability in water use of native and non-native urban trees in the greater Los Angeles area," *Urban Ecosyst.*, vol. 13, no. 4, pp. 393–414, Dec. 2010, doi: 10.1007/s11252-010-0127-6.

[152] Hwang, Y. H., Z. E. J. Yue, S. K. Ling, and H. H. V. Tan, "It's ok to be wilder: Preference for natural growth in urban green spaces in a tropical city," *Urban For. Urban Green.*, vol. 38, pp. 165–176, Feb. 2019, doi: 10.1016/j.ufug.2018.12.005.

[153] Schebella, M., D. Weber, G. Brown, and ..., "The importance of irrigated urban green space: health and recreational benefits perspectives," *Goyder Inst. Water Res. Tech. Rep. Ser.*, no. 14, 2014, [Online]. Available: http://greenspacealliance.com.au/wp-content/uploads/2020/01/The-importance-of-irrigated-urban-green-space-health-and-recreational-benefits-perspectives-Sth-Australia-doc.pdf.

[154] Ritter, L., Keith Solomon, Paul Sibley, "Sources, Pathways, and Relative Risks of Contaminants in Surface Water and Groundwater: A Perspective Prepared for the Walkerton Inquiry," *J. Toxicol. Environ. Heal. Part A*, vol. 65, no. 1, pp. 1–142, Jan. 2002, doi: 10.1080/152873902753338572.

[155] Mays, L. W., D. Koutsoyiannis, and A. N. Angelakis, "A brief history of urban water supply in antiquity," *Water Supply*, vol. 7, no. 1, pp. 1–12, Mar. 2007, doi: 10.2166/ws.2007.001.

[156] Radcliffe, J. C., "The water energy nexus in Australia – The outcome of two crises," *Water-Energy Nexus*, vol. 1, no. 1, pp. 66–85, Jun. 2018, doi: 10.1016/j.wen.2018.07.003.

[157] Grant, S. B., Fletcher, T. D., Feldman, D., Saphores, J.-D., Cook, P. L. M., Stewardson, M., Low, K., Burry, K., & Hamilton, A. J. "Adapting Urban Water Systems to a Changing Climate: Lessons from the Millennium Drought in Southeast Australia," *Environ. Sci. Technol.*, vol. 47, no. 19, pp. 10727–10734, Oct. 2013, doi: 10.1021/es400618z.

[158] Diekmann, L. O., L. C. Gray, and G. A. Baker, "Drought, water access, and urban agriculture: a case study from Silicon Valley," *Local Environ.*, vol. 22, no. 11, pp. 1394–1410, Nov. 2017, doi: 10.1080/13549839.2017.1351426.

[159] Haase, D., "Effects of urbanisation on the water balance – A long-term trajectory," *Environ. Impact Assess. Rev.*, vol. 29, no. 4, pp. 211–219, Jul. 2009, doi: 10.1016/j.eiar.2009.01.002.

[160] Gleick, P. H., "Water Use," *Annu. Rev. Environ. Resour.*, vol. 28, no. 1, pp. 275–314, Nov. 2003, doi: 10.1146/annurev.energy.28.040202.122849.

[161] Csortan, G., J. Ward, and P. Roetman, "Productivity, resource efficiency and financial savings: An investigation of the current capabilities and potential of South Australian home food gardens," *PLoS One*, vol. 15, no. 4, p. e0230232, Apr. 2020, doi: 10.1371/journal.pone.0230232.

[162] Hurd, B. H., "Water Conservation and Residential Landscapes: Household Preferences, Household Choices," *J. Agric. Resour. Econ.*, vol. 31, no. 2, pp. 173–92, 2006.

[163] Astell-Burt, T., X. Feng, S. Mavoa, H. M. Badland, and B. Giles-Corti, "Do low-income neighbourhoods have the least green space? A cross-sectional study of Australia's most populous cities," *BMC Public Health*, vol. 14, no. 1, p. 292, Dec. 2014, doi: 10.1186/1471-2458-14-292.

[164] Sahin, O., R. S. Siems, R. A. Stewart, and M. G. Porter, "Paradigm shift to enhanced water supply planning through augmented grids, scarcity pricing and adaptive factory water: A system dynamics approach," *Environ. Model. Softw.*, vol. 75, pp. 348–361, Jan. 2016, doi: 10.1016/j.envsoft.2014.05.018.

[165] Norton, B. A., A. M. Coutts, S. J. Livesley, R. J. Harris, A. M. Hunter, and N. S. G. Williams, "Planning for cooler cities: A framework to prioritise green infrastructure to mitigate high temperatures in urban landscapes," *Landsc. Urban Plan.*, vol. 134, pp. 127–138, Feb. 2015, doi: 10.1016/j.landurbplan.2014.10.018.

[166] Deitch, M. J. and S. T. Feirer, "Cumulative impacts of residential rainwater harvesting on stormwater discharge through a peri-urban drainage network," *J. Environ. Manage.*, vol. 243, pp. 127–136, Aug. 2019, doi: 10.1016/j.jenvman.2019.05.018.

[167] Sapkota, M., M. Arora, H. Malano, M. Moglia, A. Sharma, and F. Pamminger, "Understanding the impact of hybrid water supply systems on wastewater and stormwater flows," *Resour. Conserv. Recycl.*, vol. 130, pp. 82–94, Mar. 2018, doi: 10.1016/j.resconrec.2017.11.025.

[168] Kahil, T., Parkinson, S., Satoh, Y., Greve, P., Burek, P., Veldkamp, T. I. E., Burtscher, R., Byers, E., Djilali, N., Fischer, G., Krey, V., Langan, S., Riahi, K., Tramberend, S., & Wada, Y. "A Continental-Scale Hydroeconomic Model for Integrating Water-Energy-Land Nexus Solutions," *Water Resour. Res.*, vol. 54, no. 10, pp. 7511–7533, Oct. 2018, doi: 10.1029/2017WR022478.

[169] Chidya, R. C. G., W. O. Mulwafu, and S. C. T. Banda, "Water supply dynamics and quality of alternative water sources in low-income areas of Lilongwe City, Malawi," *Phys. Chem. Earth, Parts A/B/C*, vol. 93, pp. 63–75, Jun. 2016, doi: 10.1016/j.pce.2016.03.003.

[170] Birol, E., K. Karousakis, and P. Koundouri, "Using economic valuation techniques to inform water resources management: A survey and critical appraisal of available techniques and an application," *Sci. Total Environ.*, vol. 365, no. 1–3, pp. 105–122, Jul. 2006, doi: 10.1016/j.scitotenv.2006.02.032.

[171] Hurlimann, A. and S. Dolnicar, "Public acceptance and perceptions of alternative water sources: a comparative study in nine locations," *Int. J. Water Resour. Dev.*, vol. 32, no. 4, pp. 650–673, Jul. 2016, doi: 10.1080/07900627.2016.1143350.

[172] Mitchell, V. G., R. G. Mein, and T. A. McMahon, "Utilising Stormwater and Wastewater Resources in Urban Areas," *Australas. J. Water Resour.*, vol. 6, no. 1, pp. 31–43, Jan. 2002, doi: 10.1080/13241583.2002.11465208.

[173] Hatt, B. E., A. Deletic, and T. D. Fletcher, "Integrated treatment and recycling of stormwater: a review of Australian practice," *J. Environ. Manage.*, vol. 79, no. 1, pp. 102–113, Apr. 2006, doi: 10.1016/j.jenvman.2005.06.003.

[174] Low, K. G., Grant, S. B., Hamilton, A. J., Gan, K., Saphores, J., Arora, M., & Feldman, D. L. "Fighting drought with innovation: Melbourne's response to the Millennium Drought in Southeast Australia," *WIREs Water*, vol. 2, no. 4, pp. 315–328, Jul. 2015, doi: 10.1002/wat2.1087.

[175] Yannopoulos, S., I. Giannopoulou, and M. Kaiafa-Saropoulou, "Investigation of the Current Situation and Prospects for the Development of Rainwater Harvesting as a Tool to Confront Water Scarcity Worldwide," *Water*, vol. 11, no. 10, p. 2168, Oct. 2019, doi: 10.3390/w11102168.

[176] Sivanappan, R. K., "Rainwater harvesting tecniques," *Rainwater Harvest. water Manag.*, no. 14, pp. 10–15, 2006.

[177] Schwaller, C., Y. Keller, B. Helmreich, and J. E. Drewes, "Estimating the agricultural irrigation demand for planning of non-potable water reuse projects," *Agric. Water Manag.*, vol. 244, p. 106529, Feb. 2021, doi: 10.1016/j.agwat.2020.106529.

[178] Richards, S., Rao, L., Connelly, S., Raj, A., Raveendran, L., Shirin, S., Jamwal, P., & Helliwell, R. "Sustainable water resources through harvesting rainwater and the effectiveness of a low-cost water treatment," *J. Environ. Manage.*, vol. 286, p. 112223, May 2021, doi: 10.1016/j.jenvman.2021.112223.

[179] Leonard, D. and S. Gato-Trinidad, "Understanding the Effect of Household Rainwater Tanks on Hourly Residential Water Use," *J. Sustain. Water Built Environ.*, vol. 7, no. 4, p. 05021004, Nov. 2021, doi: 10.1061/JSWBAY.0000960.

[180] Musz-Pomorska, A., M. K. Widomski, and J. Gołębiowska, "Financial Sustainability of Selected Rain Water Harvesting Systems for Single-Family House under Conditions of Eastern Poland," *Sustainability*, vol. 12, no. 12, p. 4853, Jun. 2020, doi: 10.3390/su12124853.

[181] Maliva, R. G., "Low Impact Development and Rainwater Harvesting," 2020, pp. 765–825.

[182] National Research Council, *Urban Stormwater Management in the United States.* Washington, D.C.: National Academies Press, 2009.

[183] Zhang, D., R. M. Gersberg, W. J. Ng, and S. K. Tan, "Conventional and decentralized urban stormwater management: A comparison through case studies of Singapore and Berlin, Germany," *Urban Water J.*, vol. 14, no. 2, pp. 113–124, Feb. 2017, doi: 10.1080/1573062X.2015.1076488.

[184] Abbasi, T. and S. A. Abbasi, "Sources of Pollution in Rooftop Rainwater Harvesting Systems and Their Control," *Crit. Rev. Environ. Sci. Technol.*, vol. 41, no. 23, pp. 2097–2167, Dec. 2011, doi: 10.1080/10643389.2010.497438.

[185] Yazdanie, M. and K. Orehounig, "Advancing urban energy system planning and modeling approaches: Gaps and solutions in perspective," *Renew. Sustain. Energy Rev.*, vol. 137, p. 110607, Mar. 2021, doi: 10.1016/j.rser.2020.110607.

[186] Baby, S. N., C. Arrowsmith, and N. Al-Ansari, "Application of GIS for Mapping Rainwater-Harvesting Potential: Case Study Wollert, Victoria," *Engineering*, vol. 11, no. 01, pp. 14–21, 2019, doi: 10.4236/eng.2019.111002.

[187] Taylor, J. R. and S. T. Lovell, "Designing multifunctional urban agroforestry with people in mind," *Urban Agric. Reg. Food Syst.*, vol. 6, no. 1, Jan. 2021, doi: 10.1002/uar2.20016.

[188] Wang, Z., J. Li, and Y. Li, "Using Reclaimed Water for Agricultural and Landscape Irrigation in China: a Review," *Irrig. Drain.*, vol. 66, no. 5, pp. 672–686, 2017, doi: 10.1002/ird.2129.

[189] Anderson, J., "The environmental benefits of water recycling and reuse," *Water Supply*, vol. 3, no. 4, pp. 1–10, Aug. 2003, doi: 10.2166/ws.2003.0041.

[190] Ebo Yahans Amuah, E., P. Amanin-Ennin, and K. Antwi, "Irrigation water quality in Ghana and associated implications on vegetables and public health. A systematic review," *J. Hydrol.*, vol. 604, p. 127211, Jan. 2022, doi: 10.1016/j.jhydrol. 2021.127211.

[191] Srinivasan, J. T. and V. R. Reddy, "Impact of irrigation water quality on human health: A case study in India," *Ecol. Econ.*, vol. 68, no. 11, pp. 2800–2807, Sep. 2009, doi: 10.1016/j.ecolecon.2009.04.019.

[192] Shahid, Natasha,, M., Khalid, S., Murtaza, B., Anwar, H., Shah, A. H., Sardar, A., Shabbir, Z., & Niazi, N. K. "A critical analysis of wastewater use in agriculture and associated health risks in Pakistan," *Environ. Geochem. Health*, Sep. 2020, doi: 10.1007/s10653-020-00702-3.

[193] Dickin, S. K., C. J. Schuster-Wallace, M. Qadir, and K. Pizzacalla, "A Review of Health Risks and Pathways for Exposure to Wastewater Use in Agriculture," *Environ. Health Perspect.*, vol. 124, no. 7, pp. 900–909, Jul. 2016, doi: 10.1289/ ehp.1509995.

[194] Mendoza-Espinosa, L. G., A. Cabello-Pasini, V. Macias-Carranza, W. Daessle-Heuser, M. V. Orozco-Borbón, and A. L. Quintanilla-Montoya, "The effect of reclaimed wastewater on the quality and growth of grapevines," *Water Sci. Technol.*, vol. 57, no. 9, pp. 1445–1450, May 2008, doi: 10.2166/wst.2008.242.

[195] Qadir, M., Wichelns, D., Raschid-Sally, L., McCornick, P. G., Drechsel, P., Bahri, A., & Minhas, P. S. "The challenges of wastewater irrigation in developing countries," *Agric. Water Manag.*, vol. 97, no. 4, pp. 561–568, Apr. 2010, doi: 10.1016/j.agwat.2008.11.004.

Chapter 4

Effect of COVID-19 Pandemic on the Urban Heat Island: A Multi-City Case Study in India

Satyaprakash* and Pinakana Sai Deepak
Department of Civil Engineering, Sharda University, Greater Noida, UP. India

Abstract

Urban Heat Island (UHI) is defined as an urban area that is warmer than its rural counterpart due to excessive human activities. The temperature difference is significantly more during the night than during the daytime and is noticed around the year, except for during the monsoon. COVID-19 pandemic and the subsequent lockdown closed all the outside activities and forced the people to remain inside their houses. In India, the first national lockdown started on 24th March 2020 and lasted till November 2020 with phased unlock from June 2020. Further, in 2021, a second lockdown was imposed due to the second wave of the pandemic. The lockdown in the second phase started in April 2021 and lasted till July 2021, while easing some restrictions from June. Due to these lockdowns, whereas the air quality improved on account of less vehicular and other pollution, people staying in their houses increased the consumption/use of electricity for cooling/heating resulting in an increase in the Land Surface Temperature (LST) and subsequently the Urban Heat Island for the residential areas. To study the effect of lockdown on the change in UHI in metro and state capital cities of India, the authors have calculated the variation in the day UHI for 5 cities across different states in India for pre- and post-lockdowns of 2020 and 2021. The day LST has been calculated using LANDSAT 8 for 5 cities spread across different states in different climate zones of India. The LST and UHI of the cities have also been compared with the earlier values. The chapter details the different LST/UHI calculation methodologies, the use of cloud platforms such as Google Earth Engine (GEE) for the calculation of LST/UHI, and the relationship between LST, UHI with the human activities and air pollution. After analysis of 4 cities across

*Corresponding Author's Email: satya.prakash@sharda.ac.in.

different parts of India, the results indicate that there is a positive effect of the COVID-19 lockdown on the Urban Heat Islands. There was an increasing pattern of temperatures observed in the residential areas and a decreasing pattern in the industrial areas during the lockdown.

Keywords: UHI, LST, India, COVID-19, lockdown, GEE, LANDSAT

1. Introduction

Urban Heat Island (UHI) effect is the temperature increase in an urbanized area than the surrounded areas [1]. The causes for the UHI include the change in Land Use Land Cover (LULC), increased use of air conditioners, human gathering, wind blocking, etc. [2]. Land Surface Temperature (LST) is one of the key parameters in the study of land surface processes at all scales, local to global. Radiant, latent, and sensible heat at surface atmosphere interface, are all integrated with LST [3]. Several studies [4] have been done in the fields like climatology, hydrology, ecology, bio geology, global change studies, evapotranspiration, net radiation and air temperature modelling, where LST played a key role.

India recorded the first case of COVID-19 on January 30, 2020 [5]. The best way humans could think of to stop the spread of COVID-19 was lockdown. Immediate lockdown was imposed for a month from the day there was a rise in the number of cases i.e., March 24, 2020, and was extended later till May 31, 2020. Even after the lockdown was eased in May, the majority of the companies opted for work from home and because the educational institutions were closed, classes in online mode were continued. While the lockdown did benefit the environment by improving the air quality, there was an increase in the consumption/use of electricity for cooling/heating, since the population remained mostly inside and did not go out, except for essential and emergency services. Does this result in an increase in the LST and subsequently the Urban Heat Island for the residential areas? To find out and study the effect of lockdown on the change in UHI in metro and state capital cities of India, the authors have calculated the day UHI for 5 metro cities of India for pre- and post-lockdowns of 2019, 2020 and 2021.

1.1. COVID-19 and Its Effects on Environment

COVID-19, a disease which is caused by the virus named novel coronavirus SARS-COV-2, has wreaked havoc for the past couple of years. This is the third recorded outbreak of the coronavirus, while the other two named sudden acute respiratory syndrome (SARS) and middle east respiratory syndrome (MERS) spread in 2002 and 2012 respectively [6]. With over 410 million people affected and 5.8 million people deceased [7], this virus has spread over almost all countries except small isolated islands. Starting with 11th March when World Health Organisation declared SARS-CoV-2 as a pandemic [8], all the countries had only two choices to choose from, to control the number of cases as it would take some time for the vaccine to be prepared. The first one was a complete lockdown which meant to stop the spread of COVID-19 from person to person and the second one was trace-test-treat [9]. The lockdown caused a huge negative impact on global economy. A study [9] estimates that the virus had reduced the global economic growth in 2020 by -3.2%. But it also had a positive effect on the environment. As the transportation was limited to emergency and most of the industrial sector was shut down, the change in the environment was clearly visible. Air Quality Index (AQI) is generally used to indicate the severity of air pollution to the public, where high values mean more polluted and vice versa [10]. The AQI of most polluted cities reached their lowest values during the period of lockdown [11]. The NO_2 and CO_2 concentrations were reduced by 50% in China, where as NO_2 concentrations in UK were reduced by 42% during lockdown periods [11]. When the concentration levels of different pollutants in April 2019 was compared to that of lock down period, there was a decrease of 53%, 45%, 27%, 54%, and 35% in the levels of PM_{10}, $PM_{2.5}$, CO, NO_2, and SO_2 respectively [12]. Not only Air pollution, the shutdown of industries has also helped in decreasing the water pollution levels as main source of water pollution is the industrial wastes. In countries like India, Bangladesh, Malaysia, Thailand, Maldives, Indonesia the water pollution levels were drastically decreased [13]. The restricted use of vehicles also resulted in reduction of noise pollution. The noise levels of residential areas in Delhi came down to 40dB from 55dB [14].

But the pandemic also had negative impacts like increase in bio medical waste generation and municipal solid waste generation. As the pandemic dealt with high number of patients being treated in hospitals, there has been a huge rise in the bio medical waste generation [15, 16]. There might have been another negative impact of the lockdown in terms of increased UHI. This chapter tried to explore this relationship between lockdown and the increase in LST and subsequent UHI.

2. Land Surface Temperature

Land Surface Temperature is the temperature of the earth's surface and it is an important variable in the studies of land surface [17]. It helps in studying various fields which include hydrology, urban climate, environmental modelling, urban heat island [18-20]. The first method to estimate the LST used temperature sensors fixed on cars and driven on different routes. But the disadvantages of the method were expensive and time consuming. Also, the spatial interpolation method used to interpret the data for the city would not give accurate results [21]. The estimation of LST using remote sensing data was much easier than the manual method. The pros include high resolution, repetitive coverage (temporal resolution), and no manual work on site. The satellites which have TIR – Thermal Infra-Red sensors can be used to estimate LST [22]. GOES - Geostationary Operational Environmental Satellite, NOAA - AVHRR Advanced Very High Resolution Radiometer, Terra MODIS - Moderate Resolution Imaging Spectroradiometer, Aqua MODIS - Moderate Resolution Imaging Spectroradiometer, ASTER - Advanced Space borne Thermal Emission and Reflection Radiometer, Landsat-7 ETM+ Enhanced Thematic Mapper and Landsat 8 are some of the satellite data using which the estimation of LST can be done [22, 23].

It is very important to monitor the LST as it has direct correlation with Greenhouse gases in the atmosphere. With the increase in LST, the glaciers in polar regions melt down and causes an increase in the flood and sea level [24]. The most used satellite data for the estimation of LST are MODIS, and Landsat. MODIS has a temporal resolution of 1-2 days, while the one of Landsat is 16 days. While the spatial resolution of MODIS varies from 250m to 1 km for different bands, the spatial resolution of Landsat varies from 15m to 100m [17]. There are various methods to estimate LST, which are classified into three categories [25]. SEMs – Semi Empirical Methods, PBMs – Physically Based Methods and Multi-Channel TES - Temperature/Emissivity Separation methods are the three categories of methods which help in calculating LST from remote sensing data. CBEM - Classification-based emissivity method and NBEM - NDVI-based emissivity method come under Semi Empirical Methods. The retrieval of LST from Landsat is done using this method. The category of Physically-based Method includes Physics-based day/night (D/N) method and TSRM - Two-step physical retrieval method. The retrieval from TERRA MODIS is done using this method. While the Multi-channel TES methods include GBE – Grey body emissivity method, RCM - Reference channel method, TISI – Temperature independent spectral indices. The different remote sensing data which uses these methods to retrieve LST are NOAA, AVHRR, TERRA MODIS. A study [25] compared different algorithms like MWA - Mono Window Algorithm, RTE - Radiative Transfer Equation method, SCA - Single Channel Algorithm and SWA- Split Window Algorithm to estimate LST retrieval accuracy from Landsat 7 and 8. A study [26] used Temperature and emissivity separation (TES) algorithm and stated that the use of algorithm leads to acquiring the results of surface temperature and emissivity simultaneously.

2.1. Urban Heat Island

Urban Heat Island is the difference of temperature between urban environments and the surrounding areas [27]. According to UN World Urbanisation prospects 2014 [28], 66% of the total world population is set to be from urban areas. A study [29] concluded that the UHIs has an effect on global warming. As there is an increase in the use of electricity consumption in urban areas and then the use of fossil fuels, it is very important to study and monitor the so called Urban Heated Islands [30].

The causes of UHIs include [2]

1. Low evapotranspiration because of less vegetation
2. Low albedo which leads to absorption of solar radiation
3. More air pollutants
4. Blocking of wind
5. Increased use of air conditioners and GHG emitting appliances

UHI is of two different types. They are SUHI - Surface Urban Heat Islands and AUHI – Air Urban Heat Island. SUHI is basically measured using remote sensing data and they are present both during day and night but tend to be strongest during day time when the sun shines. While the AUHI is the measurement of difference in air temperature of urban areas to that of surrounding rural areas [31], it is measured using in-situ sensors [32].

SUHI – Surface Urban Heat Islands are measured using remote sensing data. The methods used to calculate this parameter are broadly classified into three categories. They include, Using LST to estimate SUHI [33-39], LST differences between Urban and Rural areas [40-46], and statistical models [47-49]. The methods which used LST to estimate UHI focused on LST variations and the relationship between LST and spatio-temporal factors of the city. The methods which used the difference of LST between urban areas and rural areas, indicated a SUHI – Surface Urban Heat Island value. With the negative value representing decrease in SUHI, the positive value represented an urban heating effect.

3. Materials and Methods

3.1. Data

Earth resources technology satellite was launched in July 1972 and later was renamed as Landsat-1. From 1972, Landsat series (1-9) monitor earth continuously and is the longest operating Earth observation satellite [50]. A joint effort of USGS – U.S Geological Survey and NASA – National Aeronautics and Space Administration, supports various applications in global change research, forestry, geology, agriculture, land cover mapping [51]. The latest satellite of the series Landsat-9 was launched in September 2021 and the data is available from 10 February 2022. Collecting 750 scenes per day, it will move into the orbit of Landsat-7, which will decommission later [52]. Landsat 8 satellite imagery has been used in this study for estimating Land Surface Temperature as it has the Landsat 8 OLI sensor with nine reflective bands having spatial resolution of 30 metres, and TIRS sensor containing two bands in Thermal Infrared Region with a spatial resolution of 100 meters. The thermal bands (Band 10 and 11) are of 100-meter resolution but are resampled and published at 30-meter resolution by USGS [53]. Landsat 8 data is available from February 2013 and is ongoing [52].

3.2. Study Area and Reasons for the Study

For the current study, 5 cities of India were taken into consideration, which are spread over different regions, North, South, East and West, of India (Figure

1). Kanpur in the north, Kolkata to the east, Surat to the west and Hyderabad to the south, the cities cover all zones of India. These cities have high population density and consistently report high air quality index. As these cities are epitome of development in India, there are a lot of industries and corporates operating. During the pandemic, all the industries and companies were shut down completely for 2 months. Even when the lockdown regulations were relaxed later, tech giants like Google, Twitter and Facebook lead the way to choose WFH for its employees. This study, calculates the Land Surface Temperature and Urban Heat Island effect of COVID-19 lockdown on these cities. The swift shift of work from industrial area to residential area can affect the temperature rise in residential areas due to increased use of air conditioners and electricity. The LST and UHI are calculated for February/March of 2019, 2020, 2021 i.e., no lock down period and April/May of 2019, 2020, 2021 i.e., lock down period.

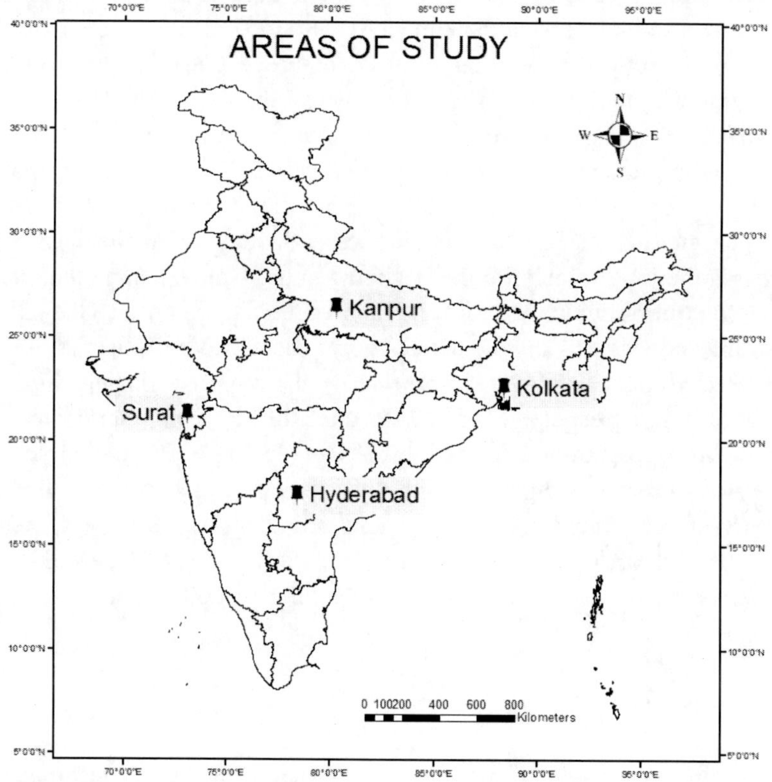

Figure 1. Areas of Study.

4. Methodology

The data for the study was processed and LST was calculated using Google Earth Engine platform using the satellite data of Landsat-8.

4.1. Google Earth Engine

Earth Engine (GEE) is a cloud computing platform which allows the analysis of free satellite images using various algorithms of digital image processing, visualization and machine learning [54]. This cloud computing platform, also known as Earth Engine, helps in decreasing the processing time and allows access to large volume of data sets at a place. Some of the datasets available in Google Earth Engine include complete Landsat series (1972 – present), MODIS images (2000 – present), Sentinel satellites (2014 – present), Advanced Land Observing Satellite (ALOS) [55]. The presence of Earth Engine code editor, which can be used to process the imagery by using JavaScript or Python is an added advantage for the users. The facets of GEE that makes it stand out from other remote sensing processing software /platforms are, its accessibility, ability to process in less time, and availability of large sets of satellite imagery. As GEE is a cloud computing platform, the required satellite imagery can be processed without downloading, which decreases the time to acquire the desired results. Unlike some other remote sensing platforms, which are paid and not available to everyone, GEE is a free to use and accessible to anyone with a google account. Various supervised and unsupervised algorithms help in processing the required imagery in a short period of time irrespective of the features of the computer/device. The processed images can be exported to the Google Drive. But the engine does come with its restrains. Specific studies like InSAR which involves the use of SAR (Synthetic Aperture Radar) – SLC (Single Look Complex images), cannot be performed in Google earth engine [56]. In the study, Google Earth Engine was used to process the images to get the desired results.

4.2. Processing

Landsat imagery is one of the most used satellite imagery to estimate land surface temperature, because of various reasons. Some of them include, its free availability, high resolution in all publicly available data of 30 meters, which are appropriate for local and small scale studies and especially availability of thermal infrared (TIR) band which is the most important band used for estimating the land surface temperature. Number of algorithms for estimation of LST using Landsat series have been proposed and used [57-65]. Algorithms like statistical mono-window, split-window, temperature/ emissivity separation, and single channel method are used depending upon the satellite data source [66]. A methodology for deriving LST from Landsat series by using the Statistical Mono-Window (SWM) algorithm [67] which is fully implemented in Google Earth Engine. The algorithm was developed by Climate Monitoring Satellite Application Facility (CM-SAF) [67]. The input data of Landsat-8 is present in Google earth engine data catalogue.

The selected days for image processing are mentioned in the Table-1. The images were selected for two periods. The first is for February or March which is the pre lock down period and the second for April or May which is the lock down period for the years 2019, 2020 and 2021. The images with high cloud cover were excluded.

The code is an open source code for generating Landsat LSTs. The user needs to provide some basic inputs like date range of observation, region of interest, NDVI flag and Landsat satellite number (4,5,7,8). The complete flow chart of the processing chain of the code is shown in Figure 2.

The results are acquired by exporting the results into Google drive and then downloading them. The links [67] to the GEE and Git repositories of the code are,

https://code.earthengine.google.com/?accept_repo=users/sofiaermida/landsat_smw_lst

https://earthengine.googlesource.com/users/sofiaermida/landsat_smw_lst

Landsat is the most used satellite imagery for Land surface temperature estimation because of its spatial resolution and presence of thermal bands. The code is an open source code for calculating the LST for the region of interest [68]. Information like the region of interest (ROI) and date of observation are required to be entered by user and the resultant images will provide the LST of the area. The image is then exported through Google drive.

Table 1. Dates of Landsat-8 Images processed

CITY NAME	Date of processed data (DD/MM/YYYY)						Satellite Used
HYDERABAD	29/3/2019	16/5/2019	15/3/2020	2/5/2020	18/3/2021	5/5/2021	Landsat-8
KANPUR	13/3/2019	30/4/2019	28/2/2020	2/5/2020	2/3/2021	5/5/2021	Landsat-8
KOLKATA	3/3/2019	6/5/2019	18/2/2020	8/5/2020	8/3/2021	25/4/2021	Landsat-8
SURAT	9/3/2019	12/5/2019	24/2/2020	14/5/2020	14/3/2021	15/4/2021	Landsat-8

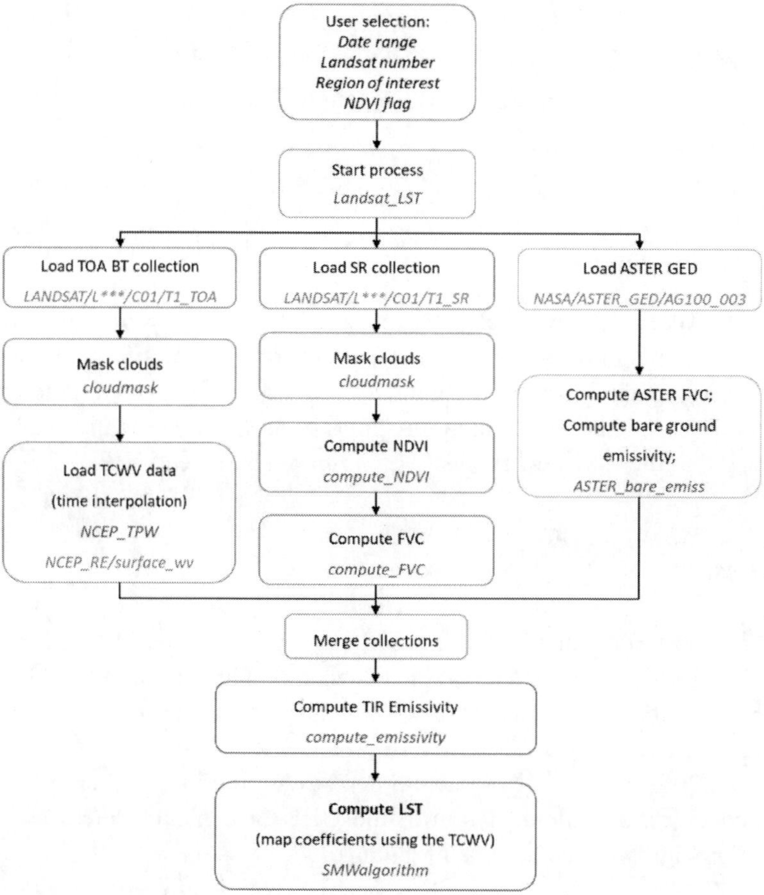

Figure 2. Process chain of the code [67].

Urban Heat Island effect of the region is calculated using the LST values acquired using Google earth engine in the study. Using the LST values, Urban Thermal Field Variance Index (UTFVI) was calculated. The formula [69] used to calculate the UTFVI is

$$\frac{T_S - T_{mean}}{T_S}$$

where,

T_s = Land Surface Temperature; T_{mean} = Mean LST

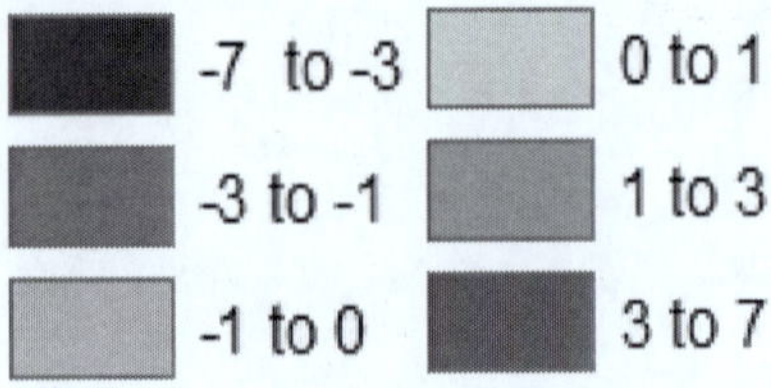

Figure 3. Colour map.

The exported images were used to calculate the UHI values with the help of raster calculator in ArcMap. The results were then classified into different zones. The Figure 3 shows the legend of the classified image UHI map, in deg Celsius. The scheme shown in Figure 2 has been followed for all the cities and dates so that the changes can easily be observed and identified.

4.3. Results

The UHI obtained maps were classified using the classes as defined in Figure 3. Profile lines were drawn on all the maps, taking into account the location of residential and industrial areas. The information about the industrial, institutional, commercial and residential areas was obtained from the ESRI-LULC 2020 map. The below section describes the profiles taken for the four selected cities, and the variation in the UHI for different dates and its co-orelation with the selected type of landuse.

4.3.1. Hyderabad

Hyderabad, the capital city of Telangana state is located at 17.3850° N, 78.4867°E, on the banks of Musi river. The city is estimated to have a population of 10 million making it the 4th most populous city in India [70]. The city is called also as 'Pearl city', 'Cybercity' and is situtated on the deccan plateau at a height of 1788 feet above sea level. The city is also home to tourist places like Ramoji film city, Charminar, Golconda fort, etc. The pollution levels in hyderabad are examined to be under the category of 'Good to moderately polluted' [71].

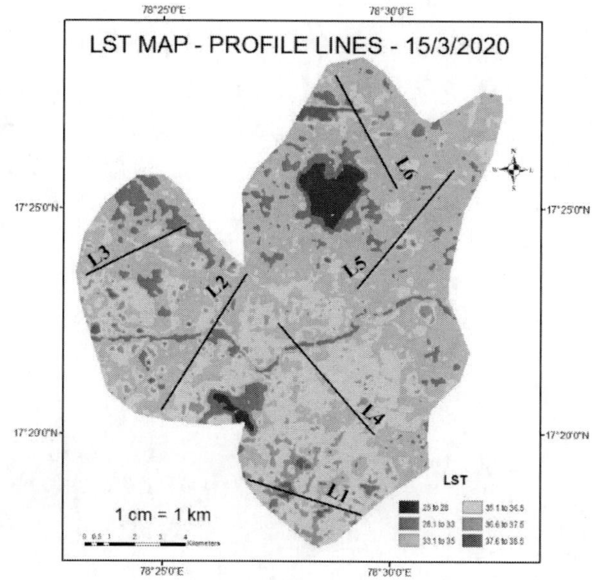

Figure 4. LST map and profile lines of Hyderabad.

4.3.2. Study Area and Profile Lines

Figure 4 shows the LST map of Hyderabad city and 6 profile lines L1-L6 were taken on the map, which passes through all the residential and industrial areas. The LST varies from a low of 25°C to 38.5°C. The profile lines were marked as shown in Figure 4 and the same were applied to other images of the year 2019, 2020 and 2021. Figure 5 shows the UHI map of Hyderabad for all the considered dates i.e., before and after lockdown of 2020 and 2021.

After processing the satellite data of Landsat-8 for the years 2019, 2020 and 2021 the maps were compiled. The resultant UHI intensity maps for the years of study have been shown in Figure-5 . The same legend colour as shown in Figure-3 has been used for all the UHI maps. The legend for profile maps is shown in Figure-6. All the profiles were plotted as shown in Figure-7, where x-axis represents the distance from the left side of the profile line and y-axis has the UHI in degree C.

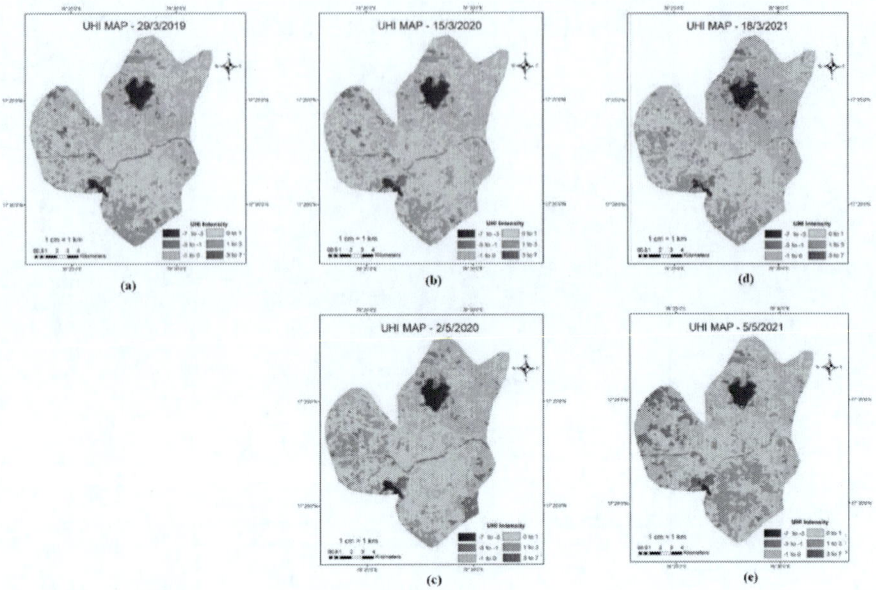

Figure 5. UHI maps of HYDERABAD for the years 2019 (left top), 2020 (centre top and bottom) and 2021 (right top and bottom).

Figure 6. Profile analysis graph lines legend.

The May 2019 image had cloud cover, hence was not included for the analysis. The Urban Heat Island Index (UHII) values were increasing from March 2020 to that of March 2021. While the west side of the area showed an increase in the UHII values from March 2019 to 2020, there was a decrease from May 2020 to 2021. The south side of the area showed a decrease of UHII values from May 2019 to 2020, and increase from 2020 to 2021. To get a better understanding of the increase and decrease of UHI for the study period in Hyderabad, the variation of UHI along the profile lines, shown in Figure 4, was plotted against distance from the starting point, and the results are compiled in Figure 7. Dashed red, green and blue lines show the UHI variations for period before the lockdown for 2019, 2020 and 2021 respectively, whereas the solid red, green and blue lines show the variations for post lockdown for the years 2019, 2020 and 2021 respectively. The combined graph for the UHI variation is shown in Figure 7.

Since there was no UHI map of Apr/May 2019, the UHI values for period before the pandemic were not studied. However, comparing the profile in Figure 7(b) and (c), which passes through the residential areas, it can be seen that the UHI values (shown by solid blue lines) of Apr/May 2021 is less than the UHI values (shown by solid green lines) of Apr/May 2020, during the pandemic. As can be observed from the profile in Figure 7(b), the UHI varies as low as 0.5°C to as high as 2.4°C. Whereas, UHI variation is almost constant of 1°C for the entire length of the profile in Figure 7(c). This supports the hypothesis that during the lockdown, since people were working from home, more heating/cooling was needed which resulted in an increase in the UHI effect in the residential areas. A reverse phenomenon was observed for the industrial areas, as can be seen from profiles in Figure 7(d)-(f). These profiles passes mainly through industrial areas and can be observed in all these profiles that UHI values of Apr/May 2021 (shown by solid blue line) is higher than the values of Apr/May 2020 (shown by solid green line). The minimum variation in these three profile is of the order of 0.2°C to a maximum of 1.5°C. As can be observed from Figure 7(f) that for the residential areas (on the right hand side of the profile), there is a marked shift in the solid blue line from the solid green line. This also conforms the second hypothesis that during the lockdown, since the industrial area was almost shutdown, except for essential services, the area recorded very less UHI as compared to when there was no lockdown.

Overall, the UHII values in Industrial area were observed to be lower in the period of lockdown i.e., May 2020. And also higher UHII values were observed in residential areas for maximum portion in the times of lockdown i.e., May 2020. The results indicate that there is an effect of lockdown on the increase of urban heat island effect in the residential areas of the city Hyderabad.

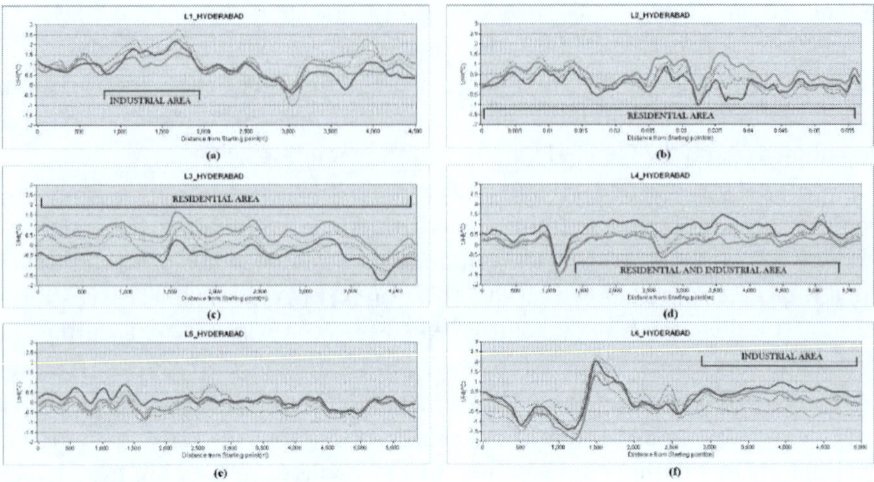

Figure 7. UHI variations for pre and post lockdown for Hyderabad along the profile lines of Figure 4.

4.3.3. Kanpur

Kanpur, a large metro city situated in state of Uttar Pradesh (UP) is located at 26.4499° N, 80.3319° E. With a population of approximately 3.1 million, Kanpur is one of the most popultaed city in UP. In 2018, Kanpur was rated as city with worst air quality by WHO [72]. A study [73] found that PM_{10} concentration is very critical in all parts of Kanpur. Situated at the banks of Ganges river, Kanpur is the industrial centre of north India. The city is also home to textile industries and is considered as "Manchester of the East" and is also called as "Leather city of the world". Since there are many and different types of industry in and around the Kanpur city, it was prundent enough to select this city and understand the effect of COVID-19 lockdown on the UHI.

4.3.4. Study Area and Profile Lines

The profile lines were marked as shown in Figure 8 and the same were applied to other images of the year 2019, 2020 and 2021.

After processing the satellite data of Landsat-8 for the years 2019, 2020 and 2021 the maps were compiled. The resultant UHI intensity (UHII) maps for the years of study have been shown in Figure 9. All the maps used the same colour scheme as shown in Figure 3.

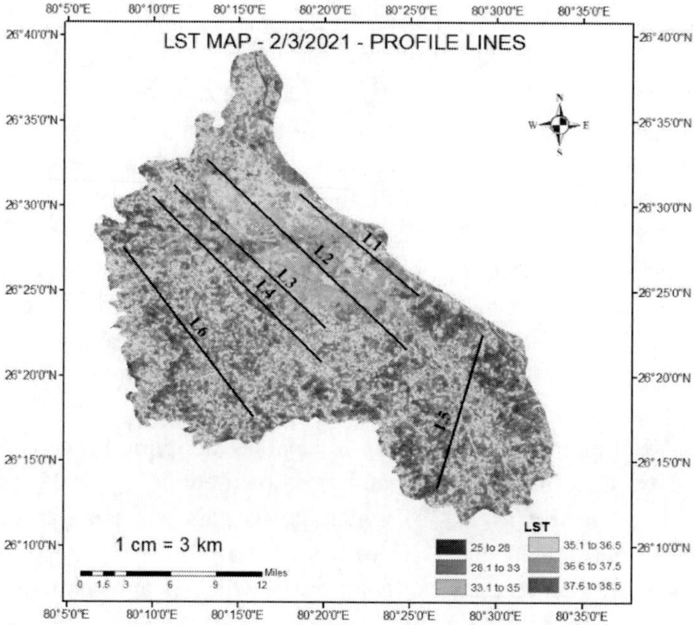

Figure 8. LST map and profile lines of Kanpur.

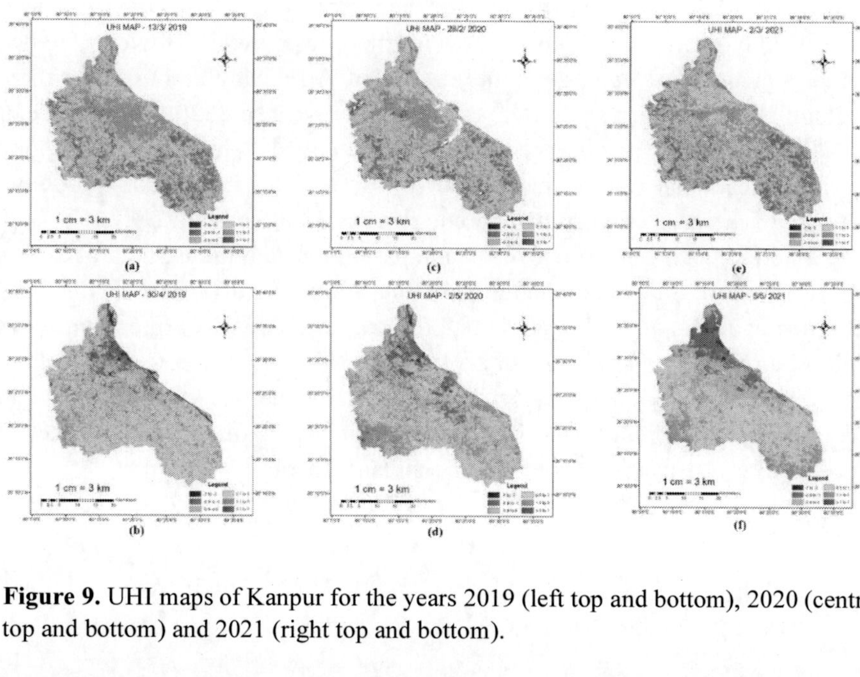

Figure 9. UHI maps of Kanpur for the years 2019 (left top and bottom), 2020 (centre top and bottom) and 2021 (right top and bottom).

Comparing all the UHII maps of pre-lockdown images (Figure 9a, c and e), it can be seen that the UHII at the central north part of the city observed a gradual decrease in the value of UHII, whereas a reverse trend is observed in the map of post-lockdown (Figures 9b, d and f). Post lockdown, the SW part of the city shows an increase in the UHI from 2019-20 whereas it decreased for the 2020-21 period. Also the northern part of the city showed a gradual decrease in the effect of UHI. To get an understiding of the effect of lockdown on UHI in the city of Kanpur, the profile analysis of 6 profiles were done. The profile lines are shown in the Figure 8 and the resultant UHII profiles are compiled in Figure 10. The colours of the profile lines are according to the legend as shown in Figure 6. The UHI values for period before the pandemic were compared to the UHI values of post pandemic.

The UHI maps for April/May of pandemic period i.e., 2020 were compared to UHI maps of pre and post pandemic i.e., 2019 and 2021. Comparing the profile (a) and (b), which mostly passes through the residential areas, it can be noticed that UHI values for months February/March (solid green) are high than April/May (solid blue) which is approximately 2°C in profile (a) and 1°C in porfile (b). In profile (b), at the end of the profile, the UHI values for 2020 is observed to be less than the that of 2019 because the profile passes through barren land, which is bound to have higher LST. Again, it can be observed in the profile (f) that the residential areas (at the right hand side of the profile) is again showing the same trend. However, when comparing the UHI values for the months of April/May amongst the three years of study, higher values were aquired in the year of 2020. An increase of as high as 1.5^0 C can be observed in the profile analysis. This city also supports the hypothesis that during the lockdown, the need of more heating/cooling because of increase in work from home resulted in increase of UHI effect.

Whereas, a part of profile (c) which passes through industrial area shows a reverse trend, albeit the change is of the order of 0.5°C. The UHI values recorded in all the three years along all the industrial area are almost same, but follows a reverse trend. A part of profile (c) which passes through residential area but is surrounded by barren land, no UHI effect found for the year of 2020, due to the presence of barren land, which has offsetted the effect of increased UHI usually observed over residential areas.

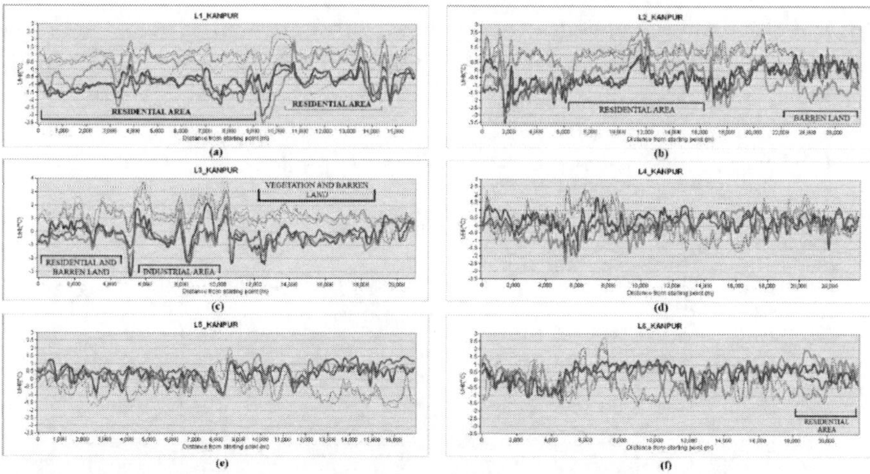

Figure 10. 4 Profile analysis for Kanpur.

4.3.5. Kolkata

Kolkata, the capital city of West Bengal is located at 22.5726°N, 88.3639°E. With 4.5 million inhibitants in the city, it is located on the eastern bank of Hooghly river. Known an "Gateway to East", suffered a loss of 2.5% of the city's GDP i.e., USD 2.1 billion due to air pollution [74]. The loss is higher then that of Mumbai and Delhi. The diseases caused by air pollution were found to be a cause of death of around 6000 people. A study [75] has been done to understand the interactive influence between Urban Heat Island and Urban Pollution Island (UPI). The study concluded a positive correlatio between UHI and UPI. It also concluded that decreasing the pollution levels might lead to decrease in the UHI levels.

The RPM and NO_2 values of the air in Kolkata have recored to cross the NAAQS standards. A study [76] in kolkata has resulted in finding that out of 17 monitoring stations, 5 were in critical category while 12 were under high category of NO_2 category. And RPM records were found to be critical in four and high in 13 stations. The major causes for the air pollution were found to be vehicular emission, industrial sources and dust particles.

4.3.6. Study Area and Profile Lines

The profile lines were marked as shown in Figure 11 and the same were applied to LST/UHI map for the year 2019, 2020 and 2021. The profile lines were selected such that it passes through both residential as well as industrial areas to see the effect of lockdown on the UHI. The image for 8th May 2020

(d) had clouds which can also be seen in the resultant map. Hence, the profile lines were selected so that they doesn't cross the cloud cover, which might lead to inaccurate results.

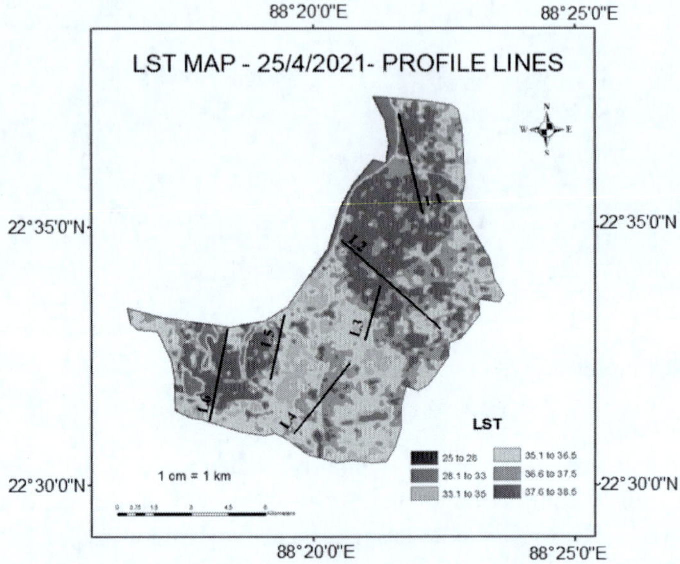

Figure 11. LST map and profile lines of Kolkata.

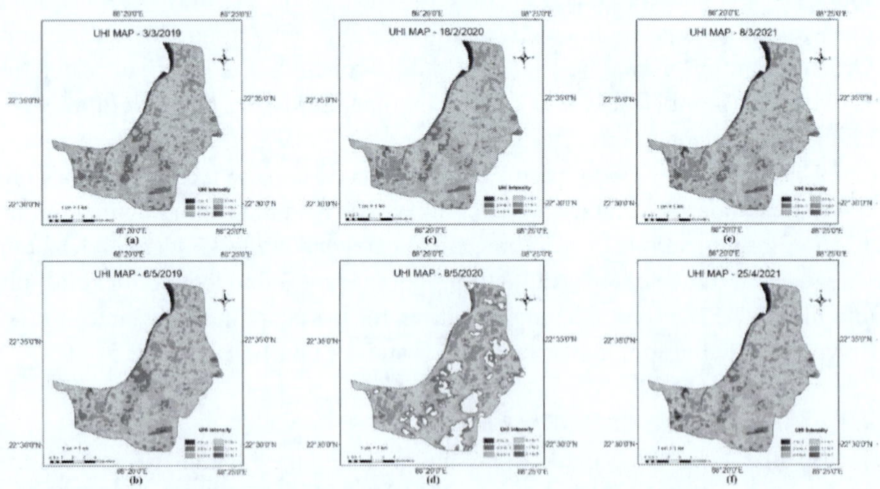

Figure 12. UHI maps of Kolkata.

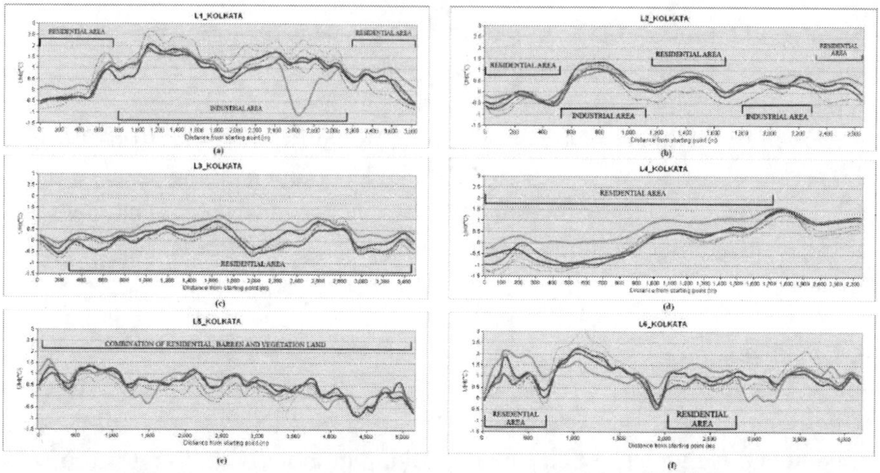

Figure 13. Profile analysis for Kolkata.

After processing the satellite data of LANDSAT-8 for the years 2019, 2020 and 2021 the maps were compiled. The resultant UHI intensity maps for the years of study have been shown in Figure 12. The legend colour map used was shown in Figure 3

There can be a decrease in the UHI noticed at the north side of the images of May 2019, 2020 and 2021 when compared to that of March 2019, 2020 and 2021. To get a clear idea of the increase and decrease among the years in Kolkata, the profile analysis for the lines showed in Figure 11 was done and the results are compiled in Figure 13.

Comparing the profiles (a), (b), (c), (d), (f) which have majority of their parts passing through residential areas, there is an increase in the UHI values for the period May 2020 (shown by solid green lines) when compared to the UHI values of other calculated periods. The increase of UHI values varies from 0.5°C to as high as 1°C. The UHI values for period of April/May in 2019 and 2021 are observed to be less than 2020 and it is also observed that both the values are almost constant through out the profiles. This supports the hypothesis that during the lockdown, there was increase in electricity consumption which lead to increased UHI effect.

While there was an increase in the UHI effect among the residential areas, a reverse phenomenon can be observed in the industrial areas. Comparing the profile lines (a), (b) where parts of them passes through industrial areas, there was a decrease in the UHI values recorded for the period of lockdown i.e., May 2020 when compared with April/May 2019 and 2021. There was a decrease of almost of 0.5°C in the UHI values. This conforms that there was a decrease in the UHI effect among the Industrial areas during the Pandemic periods as majority of the Industries were shut down.

Overall, the UHI values in Industrial area were observed to be lower in the period of lockdown i.e., May 2020 and also higher values were observed in residential areas during the same period. The results indicate that there is an effect of lockdown on the increase of urban heat island effect on regions of Kolkata.

4.3.7. Surat

Surat is a city located at 21.1702°N, 72.8311°E, in the western part of Gujarat. With population of approximately 7.7 million, Surat is second largest city in Gujarat and eighth largest city by population in India [77] . Surat is projected to be the fastest growing city in the world, with an average GDP growth rate of 9.2% over 2018 – 2035 [78]. It is also knows for diamond processing and trading. About 90% of the world's diamond supplies are cut and polished in the city. Gujarat pollution control board's annual PM 10 data of 2017 – 2018 depicts that the particle matter recorded in the city is much higher than that of National average [79]. The city experiences an average high of around 35°C and a low of 22°C with annual rainfall of 1200mm and an average relative humidity level of 53%.

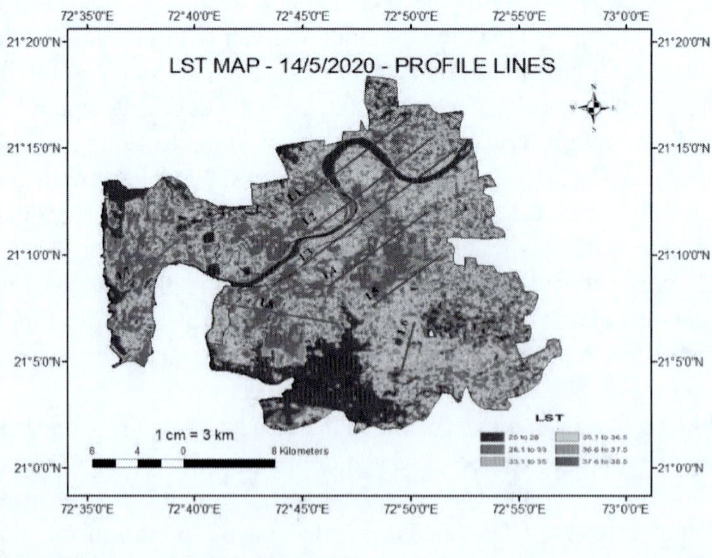

Figure 14. LST map and profile lines of Surat.

4.3.8. Study Area and Profile Lines

The profile lines were marked as shown in Figure 14 and the same were applied to other images of the year 2019, 2020 and 2021.

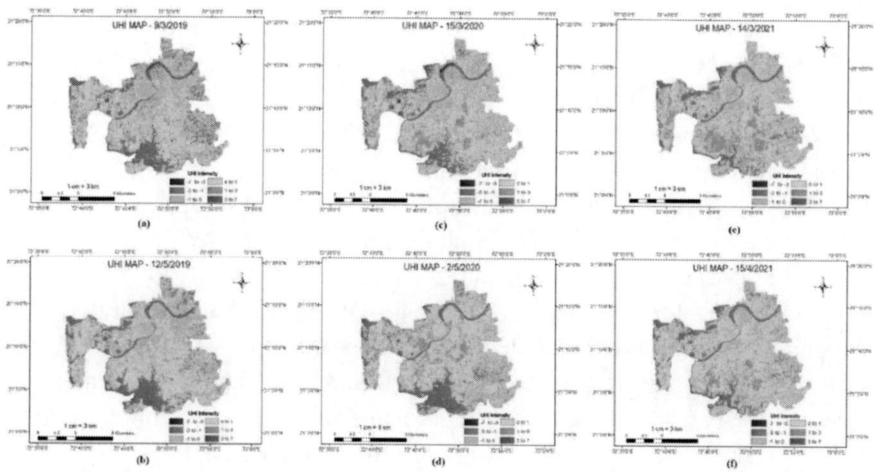

Figure 15. UHI maps of Surat.

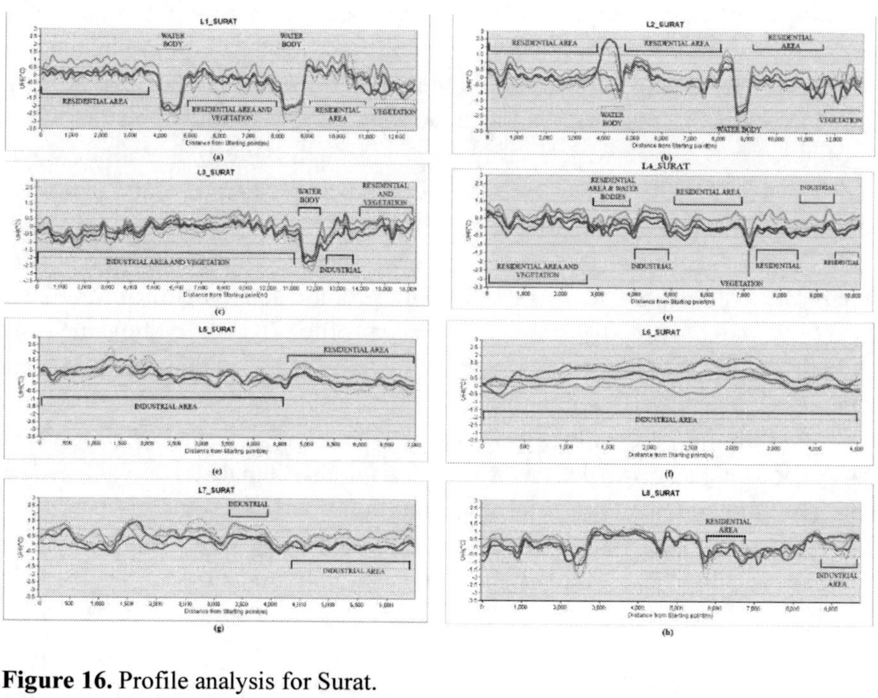

Figure 16. Profile analysis for Surat.

After processing the satellite data of Landsat-8 for the years 2019, 2020 and 2021 the maps were compiled. The resultant UHI intensity maps for the years of study have been shown in Figure 15. The colour map used was shown in Figure 3.

The part between water body i.e., in the north side is mostly residential area. The UHII values of that place recorded high in the months of May 2020 when compared to all other images. The south west part of the area also showed an increase in the UHII values from March 2019 to April 2021. There is also increase of UHII values in the middle portion of the map can be observed from 2019 to 2020 and then 2021. To get a clear idea of the increase and decrease among the years in Surat, the profile analysis for the lines showed in Figure 14 was done and the results are compiled in Figure 16.

Comparing the profiles (a), (b), (c), (d) which have majority of them passing through residential area, it is clearly evident from the pictures that the UHI values recorded in the period of April/May 2020 i.e., lockdown (shown in solid green lines) are higher than UHI values recorded in the other calculted periods. There was an increase of 0.5°C to as high as 1°C in UHI values recorded in residential areas. Profile (e), which partly covers Industrial area and residential area, showed an increased values of UHI values in residential area for the period of lockdown. This manifests that, during the lockdown as the result of people working from home, more heating/cooling was needed which then resulted in an increase in the UHI effect in the residential areas.

In contrast to the observations made among the residential areas, when profile (e) is observed which partly passes through Industrial area, there was a decrease in the UHI values recorded for the period of April/May 2020 when compared to the values of 2019 and 2021. There was a decrease in UHI values upto 0.5°C. In profile (f) which completely passes through Industrial area, there was a drop upto 2°C in the UHI values for the period of April/May 2020 when compared with 2019 and 2021. While profile (g) did not follow the same pattern, profile (h) showed an increase of UHI values in the residential areas and a decrease in UHI values among the Industrial areas.

Overall, the UHII values in Industrial area were observed to be lower in the period of lockdown i.e., May 2020. And also higher UHII values were observed in residential areas for maximum portion in the times of lockdown i.e., May 2020. The results indicate that there is an effect of lockdown on the increase of urban heat island effect on regions in Surat.

Conclusion, Future Studies and Recommendations

COVID-19 forced to impose lockdowns in India in two phases, once in March 2020 and again in April 2021. The lockdown was complete, and all the activities shifted from office/factories to home and those activities which could not be operated from home, shut down. This was true for industries and factories where complete shutdown was observed and no activity, except for some most essential and emergency services were observed. Most of the employees of IT companies, academia and others started working from home. This resulted in the reduced electricity consumption in the industrial areas for cooling and heating leading to low emissions of heat. On the other hand, as the activities shifted from office to home, there has been an increase in the consumption of electricity for heating and cooling in the residential areas. This led to increased UHI effect in the residential areas and low UHI effect in the industrial areas. With this hypothesis, this study was carried out for 4 cities spread across India. It was observed for all the cities, that the UHI effect increased for the residential areas and decreased for the industrial areas irrespective of the geographic location. The overall increase in residential areas was observed to be of the range of 0.5-2.5°C whereas for the industrial areas the decrease was almost similar. However, the variation in LST was as low as 2°C for Kolkata to a maximum of 8°C for Hyderabad. The variation in LST and UHI for the four studied cities have been compiled in Table-2.

For Hyderabad, the variation in LST was around 8°C and that for residential areas was 0.5°C more than that of the industrial areas. For Kanpur the LST variation was 10°C and residential area UHI variation was 0.5°C more. Similar was the case for Surat where LST variation was around 3°C and the difference in industrial and residential areas was 1°C. The only exception was Kolkata, where the LST varied for only 2°C and the industrial area showed an increase of 1.5°C. The lower decrease observed in the industrial areas could be attributed to the fact that not all the factories were completely shut down but factories producing essential items were operational. Whereas, for the residential areas, extended office hours and increased activities in the houses with increased occupancy all the time, increased the usage of not only heating/cooling devices but also increased use of gadgets leading to higher UHI for the residential areas. Kolkata might fall into the category of producing more essential items and hence show an increase in UHI for industrial areas as compared to the residential areas.

Table 2. Maximum and minimum of UHI and LST observed for the studied cities

	UHI VARIATION IN INDUSTRIAL AREA (°C)	UHI VARIATION IN RESIDENTIAL AREA (°C)	LST VARIATION (°C)
HYDERABAD	1 – 1.5	1.5 – 2.5	28.1 - 36.5
KANPUR	0.5 – 1	0.5 – 1.5	25 - 35
KOLKATA	0.5 – 2.5	0.5 – 1	36.6 - 38.5
SURAT	1 – 1.5	0.5 - 2	35 - 38.5

A similar study [80] has been conducted for Melbourne, Tokyo, New York City, Dublin and Oslo and concluded that the UHI pattern in cities with higher population density and urban built-up areas was more affected by the pandemic events. A follow-up to the present study may be to compare the UHI values of these cities to the pre COVID-19 times, i.e., 2018 with the UHI values of 2023, if they are similar or any change between them is observed. Although this study only refers to human activities as a possible cause of change in the UHI effect in the industrial and residential areas, there can be many other factors which generate the GHG and cause UHI effect. A comprehensive study considering the effect of climate change and human activity can also be undertaken to study the effect of other factors on the UHI change.

The study recommends to the planners and building materials supplier and researchers to start working towards making the buildings more environmentally friendly and uses materials which does not dissipate heat into the atmosphere causing UHI effect. Although lockdown decreased the UHI effect into some regions, but it increased elsewhere, keeping the overall budget of UHI the same. Only use of better materials, better architecture and town planning can reduce the ill effects of LST and UHI.

References

[1] Howard, L. (1833). The climate of London. Harvey And Darton.

[2] Nuruzzaman, Md. (2015). "Urban Heat Island: Causes, Effects and Mitigation Measures - a Review." *International Journal of Environmental Monitoring and Analysis*, 3 (2), 67. https://doi.org/10.11648/j.ijema.20150302.15.

[3] Solanky, Vijay, Sangeeta Singh, and Katiyar, S. K. (2018). "Land Surface Temperature Estimation Using Remote Sensing Data." *Hydrologic Modeling*, 343–51. https://doi.org/10.1007/978-981-10-5801-1_24.

[4] Kafer, Pamela Suelen, Silvia Beatriz Alves Rolim, Maria Lujan Iglesias, Najila Souza da Rocha, and Lucas Ribeiro Diaz. (2019). "Land Surface Temperature Retrieval by LANDSAT 8 Thermal Band: Applications of Laboratory and Field Measurements." *IEEE Journal of Selected Topics in Applied Earth Observations and Remote Sensing*, 12 (7), 2332–41. https://doi.org/10.1109/jstars.2019.2913822.

[5] Saini, G.; Deepak, P.S., (2021). Wastewater-based epidemiology for novel Coronavirus detection in wastewater. *Global J. Environ. Sci. Manage.*, 7(4): 643-658.

[6] Feehan, Jack, and Vasso Apostolopoulos. (2021). "Is COVID-19 the Worst Pandemic?" *Maturitas*, 0 (0). https://doi.org/10.1016/j.maturitas.2021.02.001.

[7] Worldometer. (2022). "Coronavirus Toll Update: Cases & Deaths by Country." *Worldometers*., 2022. https://www.worldometers.info/coronavirus/.

[8] Cucinotta, Domenico, and Maurizio Vanelli. (2020). "WHO Declares COVID-19 a Pandemic". *Acta Biomedica Atenei Parmensis*, 91 (1), 157-60. https://doi.org/10.23750/abm.v91i1.9397.

[9] Congressional Research Service. (2021). "Global Economic Effects of Covid-19." https://sgp.fas.org/crs/row/R46270.pdf.

[10] Kanchan, Amit Kumar Gorai, and Pramila Goyal. (2015). "A Review on Air Quality Indexing System." *Asian Journal of Atmospheric Environment*, 9 (2), 101–13. https://doi.org/10.5572/ajae.2015.9.2.101.

[11] Addas, Abdullah, and Ahmad Maghrabi. (2021). "The Impact of COVID-19 Lockdowns on Air Quality—a Global Review." *Sustainability*, 13 (18), 10212. https://doi.org/10.3390/su131810212.

[12] Verma, Ram Lal, and Kamyotra, J. S. (2021). "Impacts of COVID-19 on Air Quality in India." *Aerosol and Air Quality Research*, 21. https://doi.org/10.4209/aaqr.200482.

[13] Islam, S. M. Didar-Ul, Md. Bodrud-Doza, Rafid Mahmud Khan, Md. Abidul Haque, and Mohammed A. Mamun. (2020). "Exploring COVID-19 Stress and Its Factors in Bangladesh: A Perception-Based Study." *Heliyon*, 6 (7), e04399. https://doi.org/10.1016/j.heliyon.2020.e04399.

[14] Rume, Tanjena, and S. M. Didar-Ul Islam. (2020). "Environmental Effects of COVID-19 Pandemic and Potential Strategies of Sustainability." *Heliyon*, 6 (9), https://doi.org/10.1016/j.heliyon.2020.e04965.

[15] Rahman, Md Mostafizur, Md Bodrud-Doza, Mark D Griffiths, and Mohammed A Mamun. (2020). "Biomedical Waste amid COVID-19: Perspectives from Bangladesh." *The Lancet Global Health*, 8 (10), e1262. https://doi.org/10.1016/s2214-109x(20)30349-1.

[16] Saadat, Saeida, Deepak Rawtani, and Chaudhery Mustansar Hussain. (2020). "Environmental Perspective of COVID-19." *Science of the Total Environment*, 728 (138870), 138870. https://doi.org/10.1016/j.scitotenv.2020.138870.

[17] Phan, Thanh Noi, and Martin Kappas. (2018). "Application of MODIS Land Surface Temperature Data: A Systematic Literature Review and Analysis." *Journal of Applied Remote Sensing*, 12 (04), 1. https://doi.org/10.1117/1.jrs.12.041501.

[18] Haashemi, Sirous, Qihao Weng, Ali Darvishi, and Seyed Alavipanah. (2016). "Seasonal Variations of the Surface Urban Heat Island in a Semi-Arid City." *Remote Sensing*, 8 (4), 352. https://doi.org/10.3390/rs8040352.

[19] Li, Huidong, Yuyu Zhou, Xiaoma Li, Lin Meng, Xun Wang, Sha Wu, and Sahar Sodoudi. (2018). "A New Method to Quantify Surface Urban Heat Island Intensity." *Science of the Total Environment*, 624 (May), 262–72. https://doi.org/10.1016/j.scitotenv.2017.11.360.

[20] Yao, Rui, Lunche Wang, Xin Huang, Zigeng Niu, Fongfu Liu, and Qing Wang. (2017). "Temporal Trends of Surface Urban Heat Islands and Associated Determinants in Major Chinese Cities." *Science of the Total Environment*, 609 (December), 742–54. https://doi.org/10.1016/j.scitotenv.2017.07.217.

[21] Yamashita, Shuji. (1996). "Detailed Structure of Heat Island Phenomena from Moving Observations from Electric Tram-Cars in Metropolitan Tokyo." *Atmospheric Environment*, 30 (3), 429–35. https://doi.org/10.1016/1352-2310(95)00010-0.

[22] Mallick J, Kant Y, Bharath B. (2008). Estimation of land surface temperature over Delhi using Landsat-7 ETM+. *J Ind Geophys Union*., 2008, 12(3), 131-40

[23] Kumari, Babita, Mohammad Tayyab, Shahfahad, Salman, Javed Mallick, Mohd Firoz Khan, and Atiqur Rahman. (2018). "Satellite-Driven Land Surface Temperature (LST) Using Landsat 5, 7 (TM/ETM+ SLC) and Landsat 8 (OLI/TIRS) Data and Its Association with Built-up and Green Cover over Urban Delhi, India." *Remote Sensing in Earth Systems Sciences*, 1 (3-4), 63–78. https://doi.org/10.1007/s41976-018-0004-2.

[24] Solanky, V., Singh, S., Katiyar, S.K. (2018). Land Surface Temperature Estimation Using Remote Sensing Data. In: Singh, V., Yadav, S., Yadava, R. (eds) Hydrologic Modeling. *Water Science and Technology Library*, vol 81. Springer, Singapore. https://doi.org/10.1007/978-981-10-5801-1_24

[25] Sekertekin, Aliihsan, and Stefania Bonafoni. (2020). "Land Surface Temperature Retrieval from Landsat 5, 7, and 8 over Rural Areas: Assessment of Different Retrieval Algorithms and Emissivity Models and Toolbox Implementation" *Remote Sensing*, 12, no. 2, 294. https://doi.org/10.3390/rs12020294

[26] Sobrino, Jose A., Juan C. Jimenez-Munoz, Guillem Soria, Mireia Romaguera, Luis Guanter, Jose Moreno, Antonio Plaza, and Pablo Martinez. (2008). "Land Surface Emissivity Retrieval from Different VNIR and TIR Sensors." *IEEE Transactions on Geoscience and Remote Sensing*, 46 (2), 316–27. https://doi.org/10.1109/tgrs.2007.904834.

[27] Voogt, J. A. (2002). Urban Heat Island. In: I. Douglas, (Ed) Volume 3, Causes and Consequences of Global Environment Change. In: T. Munn, (Ed). *Encyclopedia of Global Environment Change*. Chichester: John Wiley & Son, Ltd.

[28] United Nations. Department Of Economic And Social Affairs. Population Division. (2015). World Urbanization Prospects: The 2014 Revision. New York: United Nations.

[29] Limas, Andrew V, Adrians Perdana, Nandhika W, and Hendy Tannady. (2014). "PEMBAHASAN MENGENAI EFEK URBAN HEAT ISLAND DAN SOLUSI ALTERNATIF BAGI KOTA JAKARTA." J@TI UNDIP: JURNAL TEKNIK INDUSTRI 9 (1). https://doi.org/10.12777/jati.9.1.29-34.

[30] Listyawati, R. N., and Prasetiyo, P. (2021). "Analysis of Urban Heat Island Phenomenon as a Global Warming Control Based on Remote Sensing in Jember Urban, Indonesia." *IOP Conference Series: Earth and Environmental Science*, 887 (1): 012002. https://doi.org/10.1088/1755-1315/887/1/012002.

[31] US EPA, OAR. (2014). "Heat Island Compendium." US EPA. June 23, 2014. https://www.epa.gov/heat-islands/heat-island-compendium.

[32] Zhou, Decheng, Jingfeng Xiao, Stefania Bonafoni, Christian Berger, Kaveh Deilami, Yuyu Zhou, Steve Frolking, Rui Yao, Zhi Qiao, and José Sobrino. (2018). "Satellite Remote Sensing of Surface Urban Heat Islands: Progress, Challenges, and Perspectives." *Remote Sensing*, 11 (1), 48. https://doi.org/10.3390/rs11010048.

[33] Song, Yang, and Changshan Wu. (2016). "Examining the Impact of Urban Biophysical Composition and Neighboring Environment on Surface Urban Heat Island Effect." *Advances in Space Research*, 57 (1), 96–109. https://doi.org/10.1016/j.asr.2015.10.036.

[34] Peng, Jian, Jinglei Jia, Yanxu Liu, Huilei Li, and Jiansheng Wu. (2018). "Seasonal Contrast of the Dominant Factors for Spatial Distribution of Land Surface Temperature in Urban Areas." *Remote Sensing of Environment*, 215 (September), 255–67. https://doi.org/10.1016/j.rse.2018.06.010.

[35] Meng, Qingyan, Linlin Zhang, Zhenhui Sun, Fei Meng, Liang Wang, and Yunxiao Sun. (2018). "Characterizing Spatial and Temporal Trends of Surface Urban Heat Island Effect in an Urban Main Built-up Area: A 12-Year Case Study in Beijing, China." *Remote Sensing of Environment*, 204 (January), 826–37. https://doi.org/10.1016/j.rse.2017.09.019.

[36] Estoque, Ronald C., and Yuji Murayama. (2017). "Monitoring Surface Urban Heat Island Formation in a Tropical Mountain City Using Landsat Data (1987–2015)." *ISPRS Journal of Photogrammetry and Remote Sensing*, 133 (November), 18–29. https://doi.org/10.1016/j.isprsjprs.2017.09.008.

[37] Wang, Shenmin, Qifang Ma, Haiyong Ding, and Hanwei Liang. (2018). "Detection of Urban Expansion and Land Surface Temperature Change Using Multi-Temporal Landsat Images." Resources, Conservation and Recycling, 128 (January): 526–34. https://doi.org/10.1016/j.resconrec.2016.05.011.

[38] "Evaluation of Urbanization Dynamics and Its Impacts on Surface Heat Islands: A Case Study of Beijing, China." (2017). *Remote Sensing*, 9 (5), 453. https://doi.org/10.3390/rs9050453.

[39] Zhang, Linlin, Qingyan Meng, Zhenhui Sun, and Yunxiao Sun. (2017). "Spatial and Temporal Analysis of the Mitigating Effects of Industrial Relocation on the Surface Urban Heat Island over China." *ISPRS International Journal of Geo-Information*, 6 (4), 121. https://doi.org/10.3390/ijgi6040121.

[40] Song, Yang, and Changshan Wu. (2016). "Examining the Impact of Urban Biophysical Composition and Neighboring Environment on Surface Urban Heat Island Effect." *Advances in Space Research*, 57 (1), 96–109. https://doi.org/10.1016/j.asr.2015.10.036.

[41] Yao, Rui, Lunche Wang, Xuan Gui, Yukun Zheng, Haoming Zhang, and Xin Huang. (2017). "Urbanization Effects on Vegetation and Surface Urban Heat Islands in China's Yangtze River Basin." *Remote Sensing*, 9 (6), 540. https://doi.org/10.3390/rs9060540.

[42] Li, Zhe, Feng Wu, Huiqiang Ma, Zhanjun Xu, and Shaohua Wang. (2022). "Spatiotemporal Evolution and Relationship between Night Time Light and Land Surface Temperature: A Case Study of Beijing, China." *Land*, 11 (4), 548. https://doi.org/10.3390/land11040548.

[43] Yang, Liuqing, Kunyong Yu, Jingwen Ai, Yanfen Liu, Wufa Yang, and Jian Liu. (2022). "Dominant Factors and Spatial Heterogeneity of Land Surface Temperatures in Urban Areas: A Case Study in Fuzhou, China." *Remote Sensing*, 14 (5), 1266. https://doi.org/10.3390/rs14051266.

[44] Yadav, Naval Kishor, Abhisek Santra, Amiya K. Samanta, Akhilesh Kumar, Shreyashi S. Mitra, and Debashis Mitra. (2021). "Understanding the Synergistic Relation between Land Surface Temperature and Different Biophysical Parameters in Haldia Industrial City of India." *Arabian Journal of Geosciences*, 14 (22), https://doi.org/10.1007/s12517-021-08769-7.

[45] Deng, Yingbin, Renrong Chen, Yichun Xie, Jianhui Xu, Ji Yang, and Wenyue Liao. (2021). "Exploring the Impacts and Temporal Variations of Different Building Roof Types on Surface Urban Heat Island." *Remote Sensing*, 13 (14), 2840. https://doi.org/10.3390/rs13142840.

[46] Estoque, Ronald C., and Yuji Murayama. (2017). "Monitoring Surface Urban Heat Island Formation in a Tropical Mountain City Using Landsat Data (1987–2015)." *ISPRS Journal of Photogrammetry and Remote Sensing*, 133 (November), 18–29. https://doi.org/10.1016/j.isprsjprs.2017.09.008.

[47] Tran, Hung, Daisuke Uchihama, Shiro Ochi, and Yoshifumi Yasuoka. (2006). "Assessment with Satellite Data of the Urban Heat Island Effects in Asian Mega Cities." *International Journal of Applied Earth Observation and Geoinformation*, 8 (1), 34–48. https://doi.org/10.1016/j.jag.2005.05.003.

[48] Shi, Hua, George Xian, Roger Auch, Kevin Gallo, and Qiang Zhou. (2021). "Urban Heat Island and Its Regional Impacts Using Remotely Sensed Thermal Data—a Review of Recent Developments and Methodology." *Land*, 10 (8), 867. https://doi.org/10.3390/land10080867.

[49] Jiang, Jie, Yandi Zhou, Xian Guo, and Tengteng Qu. (2022). "Calculation and Expression of the Urban Heat Island Indices Based on GeoSOT Grid." *Sustainability*, 14 (5), 2588. https://doi.org/10.3390/su14052588.

[50] "Developments in Landsat Land Cover Classification Methods: A Review." (2017). *Remote Sensing*, 9 (9), 967. https://doi.org/10.3390/rs9090967.

[51] "Landsat—Earth Observation Satellites." (2016). *Fact Sheet.*, https://doi.org/10.3133/fs20153081.

[52] "Landsat 9." (2019). *Fact Sheet.*, https://doi.org/10.3133/fs20193008.

[53] Sekertekin, Aliihsan, and Stefania Bonafoni. (2020). "Land Surface Temperature Retrieval from Landsat 5, 7, and 8 over Rural Areas: Assessment of Different Retrieval Algorithms and Emissivity Models and Toolbox Implementation." *Remote Sensing*, 12 (2), 294. https://doi.org/10.3390/rs12020294.

[54] "Single-Page API Reference | Google Earth Engine." n.d. Google Developers. Accessed April 29, 2022. https://developers.google.com/earth-engine/api_docs.

[55] "Earth Engine Data Catalogue." n.d. Google Developers. https://developers.google.com/earth-engine/datasets.

[56] Prakash, Satya, and Pinakana Sai Deepak. (2022). "Water Resource Management." *Advances in Geospatial Technologies*, 197–218. https://doi.org/10.4018/978-1-7998-8331-9.ch010.

[57] Vlassova, Lidia, Fernando Perez-Cabello, Hector Nieto, Pilar Martín, David Riaño, and Juan de la Riva. (2014). "Assessment of Methods for Land Surface Temperature Retrieval from Landsat-5 TM Images Applicable to Multiscale Tree-Grass Ecosystem Modeling." *Remote Sensing*, 6 (5), 4345–68. https://doi.org/10.3390/rs6054345.

[58] Wang, Fei, Zhihao Qin, Caiying Song, Lili Tu, Arnon Karnieli, and Shuhe Zhao. (2015). "An Improved Mono-Window Algorithm for Land Surface Temperature Retrieval from Landsat 8 Thermal Infrared Sensor Data." *Remote Sensing*, 7 (4), 4268–89. https://doi.org/10.3390/rs70404268.

[59] Cristóbal, J., Jiménez-Muñoz, J. C., Sobrino, J. A., Ninyerola, M. and Pons, X. (2009). "Improvements in Land Surface Temperature Retrieval from the Landsat Series Thermal Band Using Water Vapor and Air Temperature." *Journal of Geophysical Research*, 114 (D8), https://doi.org/10.1029/2008jd010616.

[60] Qin, Z., A. Karnieli, and P. Berliner. (2001). "A Mono-Window Algorithm for Retrieving Land Surface Temperature from Landsat TM Data and Its Application to the Israel-Egypt Border Region." *International Journal of Remote Sensing*, 22 (18), 3719–46. https://doi.org/10.1080/01431160010006971.

[61] Zhang, Zhaoming, and Guojin He. (2013). "Generation of Landsat Surface Temperature Product for China, 2000–2010." *International Journal of Remote Sensing*, 34 (20), 7369–75. https://doi.org/10.1080/01431161.2013.820368.

[62] Sobrino, José A., Juan C. Jiménez-Muñoz, and Leonardo Paolini. (2004). "Land Surface Temperature Retrieval from LANDSAT TM 5." *Remote Sensing of Environment*, 90 (4), 434–40. https://doi.org/10.1016/j.rse.2004.02.003.

[63] Cristóbal, Jordi, Juan Jiménez-Muñoz, Anupma Prakash, Cristian Mattar, Dražen Skoković, and José Sobrino. (2018). "An Improved Single-Channel Method to Retrieve Land Surface Temperature from the Landsat-8 Thermal Band." *Remote Sensing*, 10 (3), 431. https://doi.org/10.3390/rs10030431.

[64] Zhou, Ji, Jing Li, Lixin Zhang, Deyong Hu, and Wenfeng Zhan. (2011). "Intercomparison of Methods for Estimating Land Surface Temperature from a Landsat-5 TM Image in an Arid Region with Low Water Vapour in the Atmosphere." *International Journal of Remote Sensing*, 33 (8), 2582–2602. https://doi.org/10.1080/01431161.2011.617396.

[65] Malakar, Nabin K., Glynn C. Hulley, Simon J. Hook, Kelly Laraby, Monica Cook, and John R. Schott. (2018). "An Operational Land Surface Temperature Product for Landsat Thermal Data: Methodology and Validation." *IEEE Transactions on Geoscience and Remote Sensing*, 56 (10), 5717–35. https://doi.org/10.1109/tgrs.2018.2824828.

[66] Srivastava, Mohit Raja, and Satyaprakash, N. A. (2020). "Urban Heat Island Effect over Delhi NCR Using LANDSATTM Data." *International Journal of Global Warming*, 22 (3), 272. https://doi.org/10.1504/ijgw.2020.110865.

[67] Ermida, Sofia L., Patrícia Soares, Vasco Mantas, Frank-M. Göttsche, and Isabel F. Trigo. (2020). "Google Earth Engine Open-Source Code for Land Surface Temperature Estimation from the Landsat Series." *Remote Sensing*, 12 (9), 1471. https://doi.org/10.3390/rs12091471.

[68] "Calculating Land Surface Temperature (LST) from MODIS Dataset in Google Earth Engine (GEE)." n.d. Www.youtube.com. Accessed April 29, 2022. https://www.youtube.com/watch?v=SOSgEuqrckc.

[69] Chidi, Chhabi Lal, Ramesh Kumar Salami Magar, and Dipendra Salami Magar. (2021). "Assessment of Urban Heat Island in Kathmandu Valley (1999-2017)." *Geographical Journal of Nepal*, 14 (March), 1–20. https://doi.org/10.3126/gjn.v14i0.35544.

[70] "Hyderabad Population 2020 (Demographics, Maps, Graphs)." n.d. Worldpopulationreview.com. https://worldpopulationreview.com/world-cities/hyderabad-population.

[71] Venkateswara Rao, K., Raveendhar, N., and Swamy, A. V. V. S. (2016). Status of Air Pollution in Hyderabad City, Telangana State. International Journal of Innovative Research in Science, *Engineering and Technology*., Vol. 5, Issue 4. DOI:10.15680/IJIRSET.2016.0504023

[72] "Kanpur Population 2022 (Demographics, Maps, Graphs)." n.d. Worldpopulationreview.com. Accessed April 30, 2022. https://worldpopulationreview.com/world-cities/kanpur-population/.

[73] Singh, Karan, and Deepesh Singh. (2019). "Analysis and Mapping of Air Quality in Kanpur City." *International Journal of Engineering Technology Science and Research*, 6 (7), 89–96. http://www.ijetsr.com/images/short_pdf/1564543213_mccia705_etsr.pdf

[74] "Kolkata Population 2020 (Demographics, Maps, Graphs)." n.d. Worldpopulationreview.com. https://worldpopulationreview.com/world-cities/kolkata-population.

[75] Li, Huidong, Fred Meier, Xuhui Lee, Tirthankar Chakraborty, Junfeng Liu, Martijn Schaap, and Sahar Sodoudi. (2018). "Interaction between Urban Heat Island and Urban Pollution Island during Summer in Berlin." *Science of the Total Environment*, 636 (September), 818–28. https://doi.org/10.1016/j.scitotenv.2018.04.254.

[76] "Air Pollution and Human Health in Kolkata, India: A Case Study." (2017). Climate, 5 (4), 77. https://doi.org/10.3390/cli5040077.

[77] "Surat Population 2022 (Demographics, Maps, Graphs)." n.d. Worldpopulationreview.com. Accessed April 30, 2022. https://worldpopulationreview.com/world-cities/surat-population.

[78] "Which Cities Will Be Leading the Global Economy in 2035? | Inter American Dialogue." n.d. Globaltrends.thedialogue.org. Accessed April 30, 2022. http://globaltrends.thedialogue.org/publication/which-cities-will-be-leading-the-global-economy-in-2035/.

[79] "Ministry of Environment, Forest & Climate Change." (2019). https://cpcb.nic.in/openpdffile.php?id=UmVwb3J0RmlsZXMvMTI0M18xNjE2NTYxOTAxX21lZGlhcGhvdG8xMTgzNi5wZGY=.

[80] Wai, C. Y., Muttil, N., Tariq, M. A. U. R., Paresi, P., Nnachi, R. C., & Ng, A. W. M. (2021). Investigating the Relationship between Human Activity and the Urban Heat Island Effect in Melbourne and Four Other International Cities Impacted by COVID-19. *Sustainability*, 14(1), 378. https://doi.org/10.3390/su14010378

Chapter 5

A Holistic Approach for Outdoor Thermal Comfort Evaluation and Strategies to Reduce Urban Heat Island

Vaibhav Rai Khare[1,*], Priyanka Kumari[2] and Vaishaly[1]
[1]Environmental Design Solutions, New Delhi, India
[2]School of Environmental Science, Jawaharlal Nehru University, New Delhi, India

Abstract

Global climate change and rapid urbanization pose a threat to the environment of the planet, and it has been established that an increase in global temperature directly affects the microclimate. Urban heat island (UHI) is believed one of the most noticeable effects of climate change and rising air temperature in urban areas has been an issue of global concern since the industrial revolution. The replacement of lush greenery with high-rise structures raises urban air temperature, causing human thermal stress in urban environments. Outdoor thermal comfort is a significant aspect in determining a city's liveability as well as boosting people's health and well-being. The decreasing outdoor thermal comfort levels are one of the immediate impacts of the UHI phenomena which appeared as one of the key climatical risks to the urban inhabitants. There are over 165 different thermal comfort indices developed to represent thermal comfort based on several parameters. This chapter discusses and provides insights into the importance of outdoor thermal comfort, various thermal comfort indices, and future scope to quantify the outdoor thermal

[*]Corresponding Author's Email: vrk1007@gmail.com.

In: Urban Heat Islands Reexamined
Editors: Satya Prakash and Anne W. Ng
ISBN: 979-8-88697-215-3

environment. It has been observed that improving greenery and implementing heat mitigation strategies can enhance outdoor thermal comfort conditions. The chapter also discusses the findings of numerous outdoor thermal comfort studies to help urban designers and planners to mitigate the UHI effect and proposes a comprehensive approach for outdoor thermal comfort assessments to identify adequate methods to improve the thermal comfort level of outdoor spaces.

Keywords: urban heat island, outdoor thermal comfort, UHI mitigation strategies, UTCI, thermal comfort indices

Nomenclature

AT	Apparent Temperature
CFD	Computational Fluid Dynamics
Clo	Clothing insulation
DBT	Dry Bulb Temperature
ET	Effective Temperature
MET	Metabolic rate
MRT	Mean Radiant Temperature
OTC	Outdoor Thermal Comfort
PET	Physiological Equivalent Temperature
PMV	Predicted Mean Vote
RH	Relative Humidity
SET	Standard Effective Temperature
SR	Solar Radiation
SVF	Sky-View Factor
Ta	Air Temperature
THI	Temperature-Humidity Index
TSI	Thermal Sensation Index
TSV	Thermal Sensation Vote
UHI	Urban Heat Island
UTCI	Universal Thermal Climate Index
WBGT	Wet Bulb Globe Temperature
WBT	Wet-Bulb Temperature

1. Introduction

A major concern in today's world is the growth in the global population primarily within the metropolitan areas. Climate change and rapid urbanization are worsening the living status of people, worldwide. Presently, around 50% of the planet's population resides in urban regions and this figure is estimated to grow to 6.4 billion by 2050 as per the Department of Economic and Social Affairs, United Nations (United Nations, 2018). Staying indoors promotes an unhealthy lifestyle, which somewhat affects the occupant's health and consumes more energy. As a result, there has been an emphasis on improving the quality of outdoor spaces in the last decade.

As per a statistical valuation, outdoor leisure activities such as cycling, walking, and other similar activities, have a direct impact on physiological and psychological wellness (Niu, et al., 2015); so Outdoor Thermal Comfort (OTC) is a vital aspect of the urban environment for sustainable urban development as it encourages the active lifestyle, public and cultural activities. It is a big task for urban planners and academics to analyze the outdoor spaces by mitigating the discomfort level and making them comfortable for society.

Urbanization brings about substantial change in the natural landscape, in terms of built-up area, population density, and geometrical complexity. These modifications lead to an emission of a large quantity of waste heat and pollutants into the environment (Hart & Sailor, 2009; Kohler, et al., 2016). All these developments have an impact on the energy budget of urban areas vis-à-vis the rural areas and significantly influence urban microclimate. The complex geometry of urban areas adjusts the urban radiation budget by reflecting and absorbing short-wave radiation during the day (Hove, et al., 2015). The change in land-use land-cover patterns in urban areas has led to a thermal anomaly between rural and urban regions (Kumari, et al., 2020). This has been largely observed in all urban areas and gives rise to what we call the 'Urban Heat Island' phenomenon.

1.1. Urban Heat Island

Urban Heat Island (UHI) is a phenomenon generally linked with rapid urbanization and a great example of spontaneous climate change (Oke, 1981; Saaroni & Ziv, 2010; Alonso, et al., 2007). The increasing UHI intensity is recognized to exacerbate the urban environment, urban ecology, and human well-being. The replacement of natural surfaces with low albedo and impervious surfaces triggers a rise in urban air temperature which stimulates human thermal stress in urban areas (Yang, et al., 2016). Thermal comfort in the built environment is a critical component affecting the health, well-being, and efficiency of urban residents.

Urban Heat Island is caused due to heat absorption by dry exposed surfaces, such as pavements and roofs. It is a microscale phenomenon that depends on the intensity of solar radiation and changes diurnally and seasonally (Kumari, et al., 2021). The formation of UHI is regulated by several factors like properties of urban materials, urban geometry, and vegetation. It is now evolving as one of the climatic risks to the poor urban population in developing countries. It has adverse effects on the microclimate of the region thereby affecting the thermal condition and the outdoor comfort.

1.1.1. Cause of Urban Heat Island

There are numerous factors for the heat island phenomenon. The surface temperature in the urban areas is always greater than in rural areas because of the below reasons: 1) Human-made surfaces are usually constructed of dark material that has low albedo values; 2) warming influence of built-up areas may be the slow release of stored day-time heat from built structures; 3) Built-up areas generally have low sky-view factors which reduces the radiation loss efficiency over them leading to reduced cooling effect; 4) Maximum building materials are airtight these days so the building cannot dissipate heat by evaporation during the rainy season (Krpo, et al., 2010; Kumar & Sharma, 2020; Pandey, et al., 2009). The prominent urban characteristics contributing to heat island growth as listed below in Table 1 can be arranged into the 5 main reasons for UHI development (Kim & Brown, 2021).

Table 1. Attributes to urban heat island development

Characteristics to UHI formation	Effect on the Energy Balance
Reduced vegetation	Reduces evaporation
Use of impermeable surfaces	
Increased thermal diffusivity of urban materials	Increase heat storage
Heat trap due to building geometry	Increases long wave radiation
Low solar reflectance of materials	
Wind speed	Reduces convection
Increased energy use	Increases anthropogenic heat

1.2. Outdoor Thermal Comfort

Thermal comfort describes as the balance between heat generated by the people and losing the heat by the skin through different processes such as diffusion, evaporation, respiration, conduction, radiation, and convection (Fanger, 1972). Outdoor thermal comfort refers to people's level of comfort in an exposed outdoor area with relation to the thermal environment. Thermal comfort and heat stress are critical characteristics affecting the utilization of outdoor areas from the point of view of urban design, which may be an essential step toward improving quality of life. Many studies (Djamila, 2017; Humphreys, 1978; Linden, 2002; Schellen, et al., 2013) concentrated solely on thermal comfort in the interior environment as opposed to outdoor thermal comfort since measuring climatic factors in indoor settings is easier than in outdoor settings. However, the significance of outdoor thermal comfort is gradually being identified with shifting climate and the increase of heat stress in cities (Fong, et al., 2019).

One of the immediate consequences of the UHI phenomenon, which is rapidly developing as one of the biggest climatic risks to urban living, is deteriorating thermal comfort levels. A continual rise in temperature will aggravate heat waves, cause thermal discomfort, heat-related diseases including death. Worse, the tropics are notorious for being hot and humid all time. Because of the decrease of OTC levels, vulnerable persons, such as children, the elderly, expectant ladies, and people with health disorders, are at greater risk (Fong, et al., 2019). Thermal comfort and heat stress are key factors affecting the use of outdoor areas which can be transformed into the enhancement of quality of life.

1.2.1. Factors Governing Outdoor Thermal Comfort

Outdoor thermal comfort is a significant aspect in determining a city's livability as well as boosting people's health and well-being. Outdoor thermal comfort evaluations can be categorized into four levels such as physical, physiological, psychological, and social behaviors (Kumar & Sharma, 2020).

The physical level reflects the interface of the human body with ambient conditions like air temperature, relative humidity, air velocity, mean radiant temperature, and solar radiation.

The physiological level is the analysis of reactions of the human body to the thermal environment like skin temperature, metabolic rate, or modelling utilizing the thermal comfort indices (Fong, et al., 2019).

The psychological level reflects the occupant survey for sensation, perception, and inclination to the thermal environment. It is carried out using a questionnaire survey and discussions. The psychological answers cannot be anticipated through both above-level evaluations only.

Lastly, the level of evaluation for social behaviors includes both objective and subjective indicators obtained through interviews or surveys. It needs a large research sample since it establishes thermal comfort acceptability as a social norm (Gagge, et al., 1986; Djongyang, et al., 2010). Figure 1 depicts the broad structure for outdoor thermal comfort assessment.

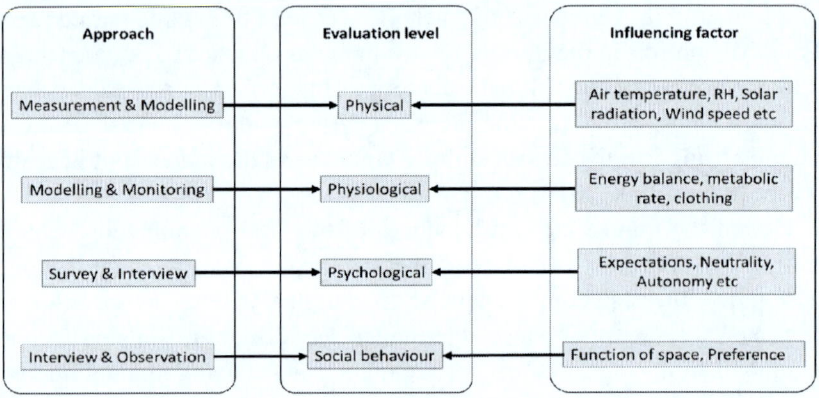

Figure 1. General framework for OTC assessment (Fong, et al., 2019).

1.3. Requirement of Outdoor Thermal Comfort Analysis

In today's fast-paced society, external spaces are just as crucial. Parks, café and restaurant sitting areas provide good exposure to sunlight and fresh air while also providing a respite from the boredom of a desk. If it isn't enough, the research aids in determining the need for additional aspects to improve outdoor thermal comfort. Below are some requirements to perform the outdoor thermal comfort analysis:

- People may determine whether the shade components will be sufficient to achieve the intended comfort conditions based on the outdoor thermal comfort study. The ability to control the air temperature in urban spaces is restricted so it can be controlled using shading to improve thermal comfort (Algeciras, et al., 2016). The shading analysis for each location can indicate the percentage of shading achieved.

- It is not viable to install mechanical active cooling systems in huge outdoor spaces. Due to a high level of ambient air infiltration and expensive installation and operating maintenance expenses, it is ineffective. As a result, the findings aid in determining which zones may benefit from an active cooling system (Mechartes, 2019). The analysis of outdoor thermal comfort will aid to determine the need for active cooling system, such as mist cooling, radiant cooling, based on practicality.
- The cooling effect of different shading scenarios on air temperature, wind velocity, and subsequent improvement to outdoor thermal comfort (Elnabawi & Hamza, 2020). Once the zone's active cooling system is constructed, comprehensive CFD wind modelling will aid in assessing the design's performance.
- It helps to predict the performance of different scenarios on urban thermal comfort using CFD modelling of wind to evaluate the performance.

2. Thermal Comfort Indices

Physical elements such as geographical location, local temperature, and the structure of the built environment impact human thermal comfort because a city has several microclimates that are adjusted at the level of different structures and areas. There are more than 165 indices have been developed to reflect thermal comfort based on a variety of criteria such as ambient temperature, subjective studies, heat transmission, and energy balance of a human body. This is the reason to develop several models to foresee the perception developed between the human and the adjacent environment from heat exchange (Blazejczyk, et al., 2012; Epstein & Moran, 2006). Several indices exist to analyze outdoor thermal conditions based on outdoor thermal conditions. These indicators are categorized in several ways. Figure 2 depicts the categorization of outdoor thermal comfort based on several models.

Numerous thermal comfort models are there to calculate outdoor human comfort (Coccolo, et al., 2016), primarily based on the energy balance between people and their surroundings, like the COMFA* model (Kenny, et al., 2009), the Standard Effective Temperature (Spagnolo & De dear, 2003) and the Index of Thermal Stress (Pearlmutter, et al., 2006). Several thermal comfort indices, listed later, have been developed to forecast observations of comfort, but many of these indices were designed to evaluate indoor comfort only (Blazejczyk, et al., 2012). Quite a few were established to calculate the thermal assessment in indoor environments (Blazejczyk, et al., 2012; Johansson, et al., 2014) and, after making some alterations, utilization in the outdoor environment can be done too. The pertinence of these indices depends upon scope, inclusiveness, transparency, usableness, and validity (de Freitas & Grigorieva, 2021). One of them is the Predicted Mean Vote (PMV), although it is recommended by the ISO 7730 (ISO, 2005) and the ASHRAE 55 (ASHRAE, 2010), these models can be challenging since it is difficult to record the dynamic features of adaption to external areas (Crawford & Stephen, 2015).

Some common outdoor thermal comfort indices (Shooshtarian, et al., 2020) include Physiological Equivalent Temperature (PET) (Hoppe, 1999), Universal Thermal Climate Index (UTCI) (Blazejczyk, et al., 2012) and Outdoor Standard Effective Temperature (OUT_SET*) (De Dear & Pickup, 2000). These indices had been exclusively intended for outdoor environments and are normally applied in outdoor thermal comfort analyses. Some analyses also utilized Predicted Mean Vote (PMV) or Adaptive Predicted Mean Vote (aPMV), but PMV's steady-state theory could make it unpredictable for varying outdoor situations (Cheng, et al., 2012).

2.1. Outdoor Standard Effective Temperature (OUT_SET*)

The effective temperature (ET*) is created on human energy balance and the two-node model (Gagge, et al., 1977). It has been improved as Standard Effective Temperature (SET*) to use for both indoor and outdoor comfort index. The SET* is a model for evaluating dry bulb temperature that compares the real-world circumstances of a situation to the effective temperature considering clo (clothing) value, metabolic rate, and 50% RH (Walls, et al., 2015). Later, in 1988, some researchers (De Dear & Pickup, 2000; Jendritzky, et al., 2012; Honjo, 2009) compared several thermal indices and reached a decision that SET* is highly capable to evaluate outdoor comfort even the convective and radiant heat exchange is limited (Chen & Matzarakis, 2018). Further, the SET* has been extended again to make specifically for outdoor conditions (Jeong, et al., 2016; Shimazaki, et al., 2011; Watanabe, et al., 2014; Xi, et al., 2012). OUT_SET* is the outdoor variation of SET* which streamlines the complex MRT conditions along with an MRT with all other variables provided as in SET*. In recent studies, many CFD models have been used in the analysis of outdoor thermal environments where the OUT_SET* is used as an output of the model (Lin, et al., 2010).

2.2. Physiological Equivalent Temperature (PET)

The Physiological Equivalent Temperature (PET) is the equivalent air temperature where the heat balance of the human body exists in typical indoor conditions (Golasi, et al., 2018). The computation of PET is established on four meteorological variables to understand the fundamental influence of these variables. It is based on a 0.9 clo thermal clothing insulation and an 80 W metabolic rate (for light activity) that must be considered as a met value (Matzarakis & Amelung, 2008; Hoppe, 2002). Additionally, it takes MRT into the account, which is equal to the air temperature, wind speed of 0.1 m/s, and the water vapor pressure to as 12 hPa (Matzarakis & Amelung, 2008). The thermal requirements of the body are then evaluated using the Munich energy balance model (Hoppe, 1993) for people, which are subsequently replaced into the energy balance equation system to produce the PET air temperature measurement. The equation for the Munich energy balance model for the individual is as follows (Girgis, et al., 2016; Hoppe, 1999; Honjo, 2009; Fong, et al., 2019):

$$M + W_o + R + C + Esk + Eres + Esw + S = 0 \ldots\ldots \quad (1)$$

Where M represents metabolic rate, Wo is mechanical work done, R is the flux of radiation, C is sensible heat, Esk is latent heat flux from the skin, Eres is latent het flux through the respiratory system, Esw is the latent heat flux through sweating, and S is the heat storage (Zero at steady state).

PET is expressed in a widely known unit, Degree Celsius (°C). Though PET is the equivalent temperature under a critical indoor condition, it applies to a broad range of actual outdoor conditions. It is used to determine a reference value among those who are unaware of human bio-meteorological conditions such as urban planners and policymakers. Table 2 demonstrates the PET ranges (Labdaoui, et al., 2021) for various levels of physiological stress at given conditions.

Table 2. Ranges of PET for different thermal perception and physiological stress

PET (°C)	Thermal perception	Grade of physiological stress
4	Very cold	Extreme cold stress
8	Cold	Strong cold stress
13	Cool	Moderate cold stress
18	Comfortable	Slight/No cold stress
23	Slightly warm	Slightly less stress
29	Warm	Moderate heat stress
35	Hot	Strong heat stress
41	Very hot	Extreme heat stress

2.3. Universal Thermal Climate Index (UTCI)

The Universal Thermal Climate Index (UTCI) is a bio meteorological tool that calculates the outside thermal conditions based on the equivalency of the physiological response anticipated by a human thermoregulation model and a state-of-the-art clothing model (Crawford & Stephen, 2015). The UTCI intends to enlighten the people of how the weather feels, considering factors such as wind, radiation, and humidity (Katić, et al., 2016). It takes all climatic factors into account such as air temperature, mean radiant temperature, relative humidity, air pressure, and wind speed. The UTCI is also described as an equivalent ambient temperature that produces the same physiological response of a reference individual as the real environment (Blazejczyk, et al., 2013; Psikuta, et al., 2012). The impetuous power behind the development of UTCI is the demand for the physiological response-based assessment model including harsh weather situations. It is a result of the world's leading comfort specialist's attempt to make an international standard of outdoor temperature sensation which is:

- The thermo-physiological implication in the entire range of heat exchange conditions of the current thermal environment.
- Effective in all climates, seasons, and scales; and
- Suitable for important applications in human biometeorology.

The UTCI index may be computed at all measurement positions for each hour of a regular year based on the relevant environmental parameters given in the weather file (Mijorski, et al., 2019). However, there is a limitation because of the fixed values used in the UTCI calculation as there might be a difference in values for humid tropic regions which may restrict the application of UTCI in tropic climate zones (Binarti, et al., 2020). The UTCI index is appropriate under the below ranges of local environmental factors (Bröde, et al., 2009):

- Air temperature between -50°C and +50°C
- Radiant temperature between 30°C below the air temperature to 70°C above the air temperature
- Wind speed at 10m above ground between 0.5 m/s and 17 m/s
- Water vapor pressure below 50 hPa or 100% relative humidity

Table 3. UTC thermal stress levels

Outdoor Comfort Level	UTCI Range (°C)	Stress Category
Level -3 and below	Below -40	Extreme cold stress
	-40 to -27	Very strong cold stress
	-27 to -13	Strong cold stress
Level -2	-13 to 0	Moderate cold stress
Level -1	0 to +9	Slight cold stress
Level 0	+9 to +26	No thermal stress
Level +1	+26 to +32	Moderate heat stress
Level +2	+32 to +38	Strong heat stress
Level +3 and above	+38 to +46	Very strong heat stress
	Above +46	Extreme heat stress

The UTCI model forecasts the thermal effect on the complete body and distinct body elements. It calculates a total of 188 points of the human body and 12 cylindrical body elements using active and passive systems (Kumar & Sharma, 2020; Fiala, et al., 2012; Blazejczyk, et al., 2013). Both passive and active system involves the calculation of heat transfer, heat convection, evaporation, and radiation (Coccolo, et al., 2016). The UTCI thermal stress levels used to measure the outdoor thermal comfort (Bröde, et al., 2013; Shooshtarian, et al., 2020; Chen, et al., 2021) are shown in Table 3.

2.4. Predictive Mean Vote (PMV)

Predicted Mean Vote (PMV) (Fanger, 1972; Chen & Ng, 2012) is the most popular indices of thermal comfort. It calculates the equilibrium thermal balance of a person and their environments established on six parameters i.e., air temperature, RH, wind velocity, MRT, clo value, and MET values by analyzing the mean thermal response of larger clusters of individuals. These variables are used in most of the physiological thermal comfort indices till yet. The PMV scale was established to define thermal discomfort which is more accurate in indoor conditions and might not be suitable for calculation of outdoor thermal comfort because of the variation in outdoor conditions (Chen & Ng, 2012; Spagnolo & De dear, 2003). The PMV model cannot give a realistic description of a human body's thermal conditions because skin temperature and sweat rate are calculated as comfort values depending upon the activity, not on climatic conditions (Gagge, et al., 1986; Fiala, et al., 2001). The PMV model was corrected for the outdoor conditions to incorporate the effects in case of extreme RH, known as the Standard Predicted Mean Vote (sPMV), and reflects the SET* in the heat balance (Gagge, et al., 1986; Walls, et al., 2015).

2.5. Thermal Sensation Index (TSI)

The Thermal Sensation Index (TSI) defines a scale ranging from 0 and 7, with 4 being the most comfortable situation (Givoni, et al., 2003). TSI was created by structured testing of participants placed in outside locations for fixed lengths of time in Japan. Subjects were asked to fill out a thermal sensation questionnaire that indicated pain, neutral, and enjoyable circumstances (Walls, et al., 2015). These tests were carried out under a variety of sun and wind conditions to assess the impact of outside environmental factors on the subject's experience. The data from the experiment was used to build an equation that expressed thermal feeling as a function of 5 factors, including the surface temperatures of nearby materials and humidity. This was then broken down into a simpler equation given below (Givoni, et al., 2003) that could be used as a forecasting formula for air temperature, solar radiation, and wind speed.

$$TSI = 1.2 + 0.115 * T_a + 0.0019 * SR - 0.3185 * v \ \ldots\ldots\ldots \quad (2)$$

Where T_a is air temperature (°C), SR is solar radiation and v is the wind speed (m/s). The equation has the R-square value of 0.87.

2.6. Apparent Temperature (AT)

Apparent temperature is specified as the perceived temperature instigated by the combined impact of air temperature, and relative humidity under normal conditions (Nasir, et al., 2013a). The measure is mostly used for the outdoor temperature. In another study, the heat index was used to define AT in °F (Nasir, et al., 2013b). The AT may be described as a representation of outside air temperature using equation (2) below:

$$AT = Ta + 0.33 * e - 0.70 * v - 4.0 \ldots\ldots.. \quad (3)$$

Where Ta is air temperature (°C), e is water vapor pressure (hPa) and v is the wind speed (m/s). The vapor pressure (e) can be calculated using the following equation (3):

$$e = 6.105 * \frac{RH}{100} * exp(17.27 * \frac{Ta}{Ta+237.7}) \ldots\ldots\ldots \quad (4)$$

2.7. Wet-Bulb Globe Temperature (WBGT)

It is a thermo-physiological indicator that is primarily utilized to assess heat stress in people exposed to direct sunshine (Honjo, 2009). WBGT is estimated using empirically derived factors depending on air and globe temperature, relative humidity, sun angle, air velocity, and solar insolation (Ghalhari, et al., 2019). The WBGT is an ISO standard (ISO 7243) that defines a testing procedure for assessing heat stress in persons who are exposed to high temperatures. The Occupational Safety and Health Administration (OSHA) and the National Institute for Occupational Safety and Health (NIOSH) use these ISO standards to assess heat exposures in the workplace (Jacklitsch, et al., 2016).

2.8. Temperature-Humidity Index (THI)

Temperature-humidity index (THI) or Discomfort index (DI) is used to measure the level of human discomfort. The discomfort level is assessed based on outdoor air temperature and RH (Din, et al., 2014). The equation for THI can be presented as below:

$$THI = 0.8 * T_a + \frac{RH * T_a}{500} \ldots\ldots\ldots \quad (5)$$

Where T_a is the air temperature (°C), and RH is the relative humidity (%). The comfort limits are defined as (Epstein & Moran, 2006):

- 100% people felt comfortable when 21 < THI < 24,
- 50% people felt comfortable when 24 < THI < 26, and
- 100% people felt uncomfortable hot when THI > 26.

2.9. Thermal Sensation Vote (TSV)

Thermal sensation vote (TSV) is proposed for outdoor thermal comfort analysis based on a large-scale survey conducted in Singapore (Yang, et al., 2013) which includes more than 2000 subjects in 13 different outdoor urban spaces. The equation to calculate TSV is as follows:

$$TSV = 0.398\, T_a + 0.023 * RH - 0.39 * v + 0.038 * MRT - 14.061 \quad (6)$$

Where Ta is the air temperature, RH is relative humidity, v is the wind speed, and MRT is the mean radiant temperature. The TSV model was adopted in the work (Ignatius, et al., 2015) to assess the outdoor thermal comfort from the influence of outdoor air temperature and wind impact. As it can be seen from the equation increases in the air temperature, RH and MRT will make people feel hot but an increase in wind speed will make people feel cooler.

There has been an increase in interest in the creation of indices for outdoor settings that may assess a population's thermal perception during the last several years (Fanger, 1972; Spagnolo & de Dear, 2003; Chen & Ng, 2012; Johansson, et al., 2014; Rose, et al., 2010; Davis, et al., 2006). As a result, a tendency toward the development of empirical indicators (Johansson, et al., 2014) has been noticed. The most widely used approach is a field survey in which interviewees are asked to rate the external temperature environment on a subjective scale (Canan, et al., 2019). Simultaneously, micrometeorological variables are monitored, enabling the development of models that may predict human heat perception using multiple regressions.

Table 4. Summary of some common thermal comfort indices used

Indices	Parameters considered	Indication	Applicability	References
Apparent temperature	Air temperature, RH under normal conditions	Comparable thermal sensation indication	Applicable for outdoor conditions	(Pigliautile, et al., 2020; Spagnolo & De dear, 2003)
Standard Predicted Mean Vote (SPMV)	Adjusted PMV to include more extreme humidity	Quantifies discomfort on a scale between -3 to +3	Doesn't very accurate for extreme conditions (outdoor).	(Cheng, et al., 2012; Golasi, et al., 2018)
Physiologically Equivalent Temperature (PET)	Four meteorological variables, clothing, activity level	Air temperature required to reproduce a comfortable indoor setting	Assume RH as 50% which changes in outdoor situations thus less accurate	(Lin, et al., 2010; Ziaul & Pal, 2019; Johansson, et al., 2014)
Outdoor Standard Effective Temperature (OUT_SET*)	Derived from SET with simplified MRT. Assume activity and clothing for outdoor usage	Compare individual comfort to a reference environment	Considers the thermos-regulation of human body thus more accurate for extreme conditions (outdoors)	(De Dear & Pickup, 2000; Davis, et al., 2006; Rose, et al., 2010)
Thermal Sensation Index (TSI)	Air temperature, solar radiation, wind speed	Quantifies discomfort on a scale between 1-7	Applicable for outdoor conditions	(Givoni, et al., 2003; Walls, et al., 2015)
Universal Thermal Climate Index (UTCI)	Air temperature, Mean temperature, wind speed, water vapour, relative humidity, clothing	Indication of physiological thermal stress under a wide range of conditions and climates	Applicable for outdoor conditions	(Golasi, et al., 2018; Hoppe, 2002; Lai, et al., 2014)

Table 4 encapsulates some of the frequently used thermal comfort indices which are applied in thermal outdoor areas.

3. The Interrelation between Outdoor Thermal Comfort and Building Energy Use

Outdoor thermal comfort and building energy use are associated in a complex way that involves both urban microclimatic and behavioral factors (Aktas, et al., 2020). Figure 2 demonstrates the connections between thermal comfort and building energy use. Buildings, which are open and have transient systems, exchange their energy with their surroundings (Zhu, et al., 2018). Overheating or undercooling of the outdoor area adds to the heat transfer amongst outdoor and indoor, thus affecting indoor and outdoor human behaviors (Yang & Wang, 2013). If the outdoor area is not comfortable then it increases comfort need in the built environment then one would anticipate people to act in response to discomfort, and hence human behavior is connected to thermal comfort and energy use (Ahmed, et al., 2021). This type of framework needs to be addressed thoroughly. People would be ready to give their time out of the buildings if the outdoor temperature is in the range, hence lowering the use of energy-consuming devices in the buildings, mainly air-conditioners (Rijal, et al., 2007). In this perception, social behavior response to the outdoor thermal environment can influence the energy consumption of buildings (O'Brien & Gunay, 2014).

In addition, building microclimate exchanges energy with surroundings, convection, radiation, and conduction between buildings and their environments. The outdoor thermal environment interacts with the building microclimate to which one is exposed when working or staying indoors (Allegrini, et al., 2012).

A study focused on the microclimate of a courtyard house that has a shape and structure typical of vernacular buildings (Du, et al., 2014). The courtyard house has indoor space, semi-outdoor space, and outdoor space; thus, it creates a broad building microclimate. Thermal parameters were investigated in all the indoor as well as outdoor areas, and then the thermal performance of the building microclimate was evaluated through thermal and CFD simulations. The findings of this study provide that the building has its microclimate environment, which interacts with the outdoor climate (Asadi, et al., 2016; Albatayneh, et al., 2018; Du, et al., 2019). The results stress the importance of coupling outdoor thermal comfort and building energy consumption.

4. Urban Characteristics Impacting Outdoor Thermal Comfort

There are various approaches to alleviate the UHI effect, putting a positive influence on both local and worldwide climate. As the study of outdoor thermal comfort and UHI is interrelated, mitigation measures in UHI can implicitly affect thermal comfort in the outdoors also. Therefore, this section focuses on urban characteristics like urban geometry, greening strategies, high albedo materials, and water bodies. First, the effect of these strategies on air temperature, wind speed, radiation, and thermal comfort has been discussed.

4.1. Urban Geometry

Urban geometry plays a crucial role in enhancing thermal comfort in the outdoor area. In particular, shading by building becomes significant in mitigating heat stress in urban areas (Andreou, 2013). In high-density towns, the understanding of how urban layout influences the outdoor thermal environment is limited. It results in inefficient urban planning, which may worsen heat discomfort in densely populated places. The following parameters are commonly used to quantify urban geometry: Factors affecting sky views, orientations, and the height-to-width(H/W) ratio (Ali-Toudert & Mayer, 2006; Lai, et al., 2017).

The H/W ratio, which is specified as the proportion of building height to street width, is a key indicator of an urban canyon's openness. In general, a greater H/W number indicates less openness and vice versa. The sky-view factor (SVF) is a dimensionless integer between zero and one that represents the amount of clear sky visible from a particular position (Lai, et al., 2017). A lower SVF implies that urban components such as buildings and trees are obstructing the sun and sky, causing a small amount of radiation from the sun, and increased long-wave radiation from the surface (Oke, 1988; Lai, et al., 2019).

Shading is a highly effective strategy for countering daytime heat stress. The source of shading (e.g., buildings or trees) is more important than the level of shading provided (Emmanuel, et al., 2007; Lin, et al., 2010).

4.1.1. Impact of Urban Geometry on Radiations

Reduced exposure to solar radiation is a feature of compact urban environments with a low sky view factor or high H/W, resulting in enhanced

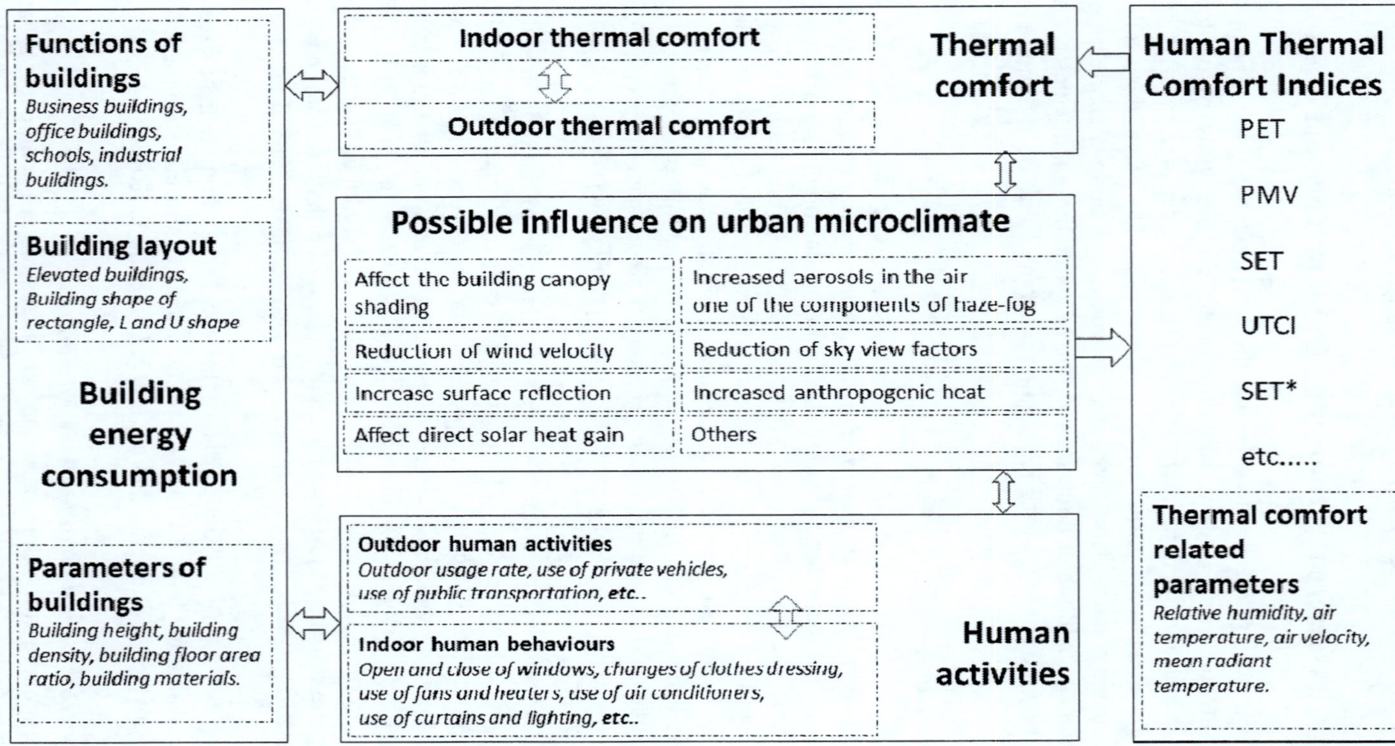

Figure 2. Brief connection amongst outdoor thermal comfort and building energy use (Li & Liu, 2020).

thermal comfort in hot regions (Tan, et al., 2013; Charalampopoulos, et al., 2013). In a school in Taiwan, the thermal comfort was evaluated for different spaces with different SVF, and it was found out that the PET was decreased while reduction in SVF (Shih, et al., 2017). Similar results were also observed in Malaysia (Qaid, et al., 2016), Bangladesh (Kakon, et al., 2009), and Huawei, Taiwan (Lin, et al., 2010).

Besides the reduced sun exposure, radiation levels in compact locations are also lower than in wide spaces. The direct correlation between SVF and MRT has been established in several studies (Wang, et al., 2016; Lai, et al., 2019).

A study (Taleghani, 2018) showed that the orientation also significantly influences the amount of radiation it receives and Intermediate orientations such as Northeast Southwest and Northwest Southeast have been explored less frequently than East-West and North-South orientations (Lau, et al., 2016). Even though the discomfort interval of these orientations was longer than that of a North-South direction with the same H/W ratio, it was observed that the intermediately orientated pathways were always substantially shaded, providing people with an alternate (Lai, et al., 2019; Chatzidimitriou & Yannas, 2017). While considering a winter scenario, the Northeast Southwest and Northwest Southeast orientations may be a reasonable compromise because they provide more sun coverage than the North-South direction.

4.1.2. Impact of Urban Geometry on Winds

When the wind encounters a structure, it gradually decreases. The more open and urban form is, the more wind it is exposed to. A parametric study using CFD (Yuan & Ng, 2012) develops a wind speed categorization based on the outdoor thermal comfort on the pedestrian level to calculate the natural ventilation in street canyons and it was concluded that lowering the site exposure ratio supports enhance ventilation. On the other hand, a study (Berardi & Wang, 2016) found that when new buildings were constructed, the wind speed increased at the building corners. The flow is directed into the canyon at the surface level if the street is orientated in the same direction as the wind. Making the street alignment parallel to the sea breeze (thus normal to the coast) is an excellent ventilation approach in coastal cities.

4.1.3. Impact of Urban Geometry on Air Temperature

Direct solar radiation in a compact area makes the space cooler compared to the open area. A study (Berardi & Wang, 2016) found that the addition of new constructions reduced the outdoor air temperature by up to 1°C during the daytime because a positive correlation exists between SVF and air temperature. This correlation is also gets confirmed through field measurement and simulation studies (Wang, et al., 2016; Cheng, et al., 2012). These studies concluded that an increase of about 60% in SVF leads to an air temperature difference of up to 1.5°C. However, the intensity of anthropogenic heat is often higher in urban areas rather than rural areas so the actual relationship between SVF and air temperature can be negative (Horrison & Amirtham, 2016).

4.1.4. Impact of Urban Geometry on Thermal Comfort

The fundamental impact of urban geometry on thermal comfort is that it alters radiative and convective heat exchange in urban open areas. Solar radiation is typically blocked by dense urban architecture, which also affects wind speed (Yang, et al., 2020).

In comparison to open places, the levels of radiation, wind speed, and air temperature in dense urban environments are often lower (Li, et al., 2022). Reduced radiation and air temperature are beneficial in hot climates for producing thermally acceptable conditions. In the summer, though, a reduction in wind speed exacerbates the urban thermal climate. Thermal indicators like PET can be used to assess these combined effects (Lai, et al., 2019; Andreou, 2013). According to quite a few studies (Ali & Patnaik, 2018; Matzarakis & Amelung, 2008; Mayer, et al., 2008), PET values in dense areas were smaller than in open areas demonstrating that the effect of shade balances the consequence of reduced wind and this result was backed by parametric analysis (Cheung & Jim, 2018; Kántor, et al., 2018). Dense urban forms show improved thermal conditions throughout summer but are inconvenient in winters (Jamei & Rajagopalan, 2017; Lai, et al., 2019). This conclusion was backed by a parametric analysis (Andreou, 2013) in which he found that PET difference between shaded and exposed area was near to 10°C, while the PET difference between wind speed was only 6°C. A trade-off between the summer and winter times should be addressed while developing urban morphology for outdoor thermal comfort, particularly in moderate areas (Chatzidimitriou & Yannas, 2015).

4.2. Greening Strategies

Vegetation in urban open areas can contribute to high-quality urban life in a variety of ways. Vegetation decreases heat in three ways (Panferov, et al.,

2001): 1) by evaporation and transpiration; 2) through greater albedo of leaves that reflect the light, and 3) by blocking solar radiation.

Trees efficiently diminish thermal radiation in urban open areas by eradicating the incoming short wave solar radiation with the help of reflection and absorption (Wong, et al., 2011). It has been found out from several studies that only by repositioning tree locations in road parking (Ballinas & Barradas, 2016), an improvement of 77% in outdoor thermal comfort can be attained (Milošević, et al., 2017). Radiation can be further minimized by implanting trees with bigger crowns (Bowler, et al., 2010). A study concluded (Wang & Akbari, 2016) that by increasing the tree coverage and using trees with bigger crowns, mean radiation temperature was significantly reduced.

The roughness of the urban surface is exacerbated by trees, which creates a drag on airflow. Porous trees have a distinct influence on limiting urban airflow than solid structures. Buildings produce a large pressure differential between the windward and leeward directions, but trees create a considerably lower pressure difference due to their permeable nature. Consequently, the wind becomes faster around building edges and roofs, while trees generate smooth wind speed fluctuations (Lai, et al., 2019). Trees in the street have various environmental benefits for urban people. A study has been performed in the temperate climate of Melbourne, Australia in which three east-west streets during heatwaves are measured (Coutts, et al., 2008). They used the UTCI for the estimation of outdoor thermal comfort within the tree canopies and reported that the tree cut down the UTCI by 6°C and the maximum air temperature drop was 1.6°C.

In addition to preventing short-wave radiation, vegetation suppresses long-wave radiation by lowering surface temperature through transpiration. The surface temperature of a green turf (34.0°C) remains significantly smaller than the concrete (45.4°C) (Chatzidimitriou & Yannas, 2017). Planting plants on rooftops can also help to lower air temperature but the cooling impact on pedestrian level depends upon the roof height of the building. Green roofs were shown to lower the average ambient air temperature by 0.4 to 3°C when applied on a city scale (Santamouris, 2014). Green roofs can have a cooling impact on pedestrians when installed on buildings with a height of fewer than 10 meters, according to a study done in Singapore (Wong, et al., 2003).

Green infrastructures (i.e., green roof, green wall, and urban trees and park) cool the atmosphere through evapotranspiration and shading, reduce the amount of absorbed solar radiation, and lower the convective and radiative heat fluxes (Koc, et al., 2018). The cooling benefits of green infrastructure depend on the vegetation type, height, thickness, availability of moisture, foliage density, and retention level in a green infrastructure system (Jamei, et al., 2020).

4.3. High Albedo Materials

Reflective surface alters the atmosphere in respect of radiation and air temperature (Akbari & Matthews, 2012). Natural or artificial materials with high reflectivity can be used to the façade, roofs of buildings, and the pavements of urban areas, to mitigate the UHI impact (BEE, 2017).

Cool materials lower surface temperatures and diminish convective heat transfer from the surface to the air, resulting in cooler air temperatures than typical materials. Reflective and white roofs provide many advantages for urban building thermal comfort. Several studies (Taha, et al., 1992; Akbari & Konopacki, 2005; Khare, et al., 2021) have demonstrated the benefit of white roofs on summertime interior cooling energy savings. Others (Garg, et al., 2016; Shickman & Rogers, 2020; Wang, et al., 2015) found that cool roofs had less thermal stress (and hence a longer lifespan) than other roof types. White roofs reflect most of the sun's short-wave energy into the sky, so the roof's heat absorption is reduced (Fayad, et al., 2021). Roofs get the greatest amount of solar radiation throughout the summer due to the high altitude of the sun (Taleghani, 2018). The albedo of the roof surface determines the proportion of re-radiating the sun. The long-wave radiation in space can be lowered by reducing the surface temperature. Although a reflecting surface doesn't increase the surface temperature, numerous simulation studies imply that more reflected solar radiation on the human body causes higher pedestrian discomfort (Yang, et al., 2016; Akbari & Matthews, 2012).

A study in a dense urban area in Ho Chi Minh, Vietnam reported that the application of a light color roof reduces the air temperature at the roof levels by 0.07°C (Huynh & Eckert, 2012). In addition to the simulation studies, a sensitivity test conducted in the field by (Rosso, et al., 2016) and established that the asphalt was less preferred than the material with higher albedo in terms of thermal comfort. It was also discovered that different varieties of gravel with different albedo have insignificant differences in heat perception (Falasca, et al., 2019). Aside from thermal preferences, it's worth noting that when albedo increased, people's visual perception degraded because of glare (Hinkle, 2016).

4.4. Water Bodies

The capacity of water bodies to cool the air through evaporation has the greatest influence on the urban thermal environment (Oke, 1988). Furthermore, the surrounding buildings and grounds temperature come down due to the high thermal capacity of water bodies. Water bodies provide a better thermal gradient for convective heat transfer and release less radiation due to the low temperature of the water (Kalinowska, 2019). The thermal capacity of water is higher than pavement materials like concrete, asphalt, granite, etc. (Chatzidimitriou & Yannas, 2017). As a result, water shows a less temperature increment compared to traditional structures and pavement materials for the same quantity of solar energy. Water bodies may be used as heat sinks in metropolitan areas (Lin, et al., 2020) as the evaporation of water absorbs heat from the surrounding environment. Furthermore, convective heat transfer between the ambient air and the water surface cools the surrounding air. A water body can drop the adjacent temperature by 2-6°C (Khare, et al., 2021) which led to a drop in air temperature in urban areas. An assessment report (Manteghi, et al., 2015) addressed that a decrease of 1-2°C had been found in surrounding temperatures with the presence of adjacent water bodies, even for the severe weather conditions.

Since water bodies are not able to prevent direct solar radiation, the enhancement in thermal comfort using them is always less than adding plantations and adjusting the urban geometry (Dincer & Rosen, 2013). In a study, PET gets reduced due to the presence of a water fountain was found only 1.4°C, which is considerably low (Chatzidimitriou & Yannas, 2015). Furthermore, the cooling impact of water bodies varies with wind conditions and distance. In hot weather, high humidity can prevent perspiration from evaporating heat from the human body. As a result, the trade-off between lower air temperature and higher humidity caused by vegetation and water bodies must be quantified in terms of human comfort (Syafii, et al., 2016).

5. Case Studies

This section discusses some case studies which shows the impact on urban heat island and outdoor thermal comfort while applying the reduction strategies considered in the chapter.

5.1. Case Study-1: A School in Hot-Humid Climate of China

One study (Zhang, et al., 2017) focused on outdoor thermal comfort and energy efficiency in one of the school campuses situated in China, where building density, building properties, building usage, and human activities are different than city level. The simulation has been performed varying building height, façade models, building roof models, and landscapes design. An integrated design approach was established combining the most effective strategy in each phase, including the mid-rise residential complex, green façade, cool roof, plantation, and grass pavement. The method proved to be effective and integrated design reduced the outdoor discomfort time by 25% and the building demand by 5%. The main implications from the study are as follows:

- Urban geometry influences the wind pattern, especially the spaces close to the buildings, thus mitigating heat stress.
- Building layout and orientation affect the outdoor environment. The wind pattern and solar radiation change/block and mitigate the heat stress during the hottest time period.
- A cool roof, green roof, and green façade decrease building cooling demand; however, their outdoor spatial influence is limited because the building façade only has a measurable impact on the spaces close to the buildings.
- If compared with all the options, plantation of the tree is found to be the most effective strategy to reduce the discomfort hours, and around 2.6% cooling demand decreased.
- The thermal comfort variation in the large areas is small but spaces near the building are affected more in terms of shading. A suggestion has been made for the outdoor activities on the playground, i.e., the outdoor activities should be avoided between 11:00 am to 05:00 pm since it is difficult to mitigate the heat stress.

This case study showed that it is critical to use an integrated design approach that considers outdoor thermal comfort and energy consumption both in the early-stage design of the building. One of the limitations of the study is that it does not provide simulation for all possible scenarios in the different design stages since the inputs might differ in terms of cities and areas.

5.2. Case Study-2: A Public Space in Hot and Dry Climate of Baghdad

The study (Salman & Saleem, 2021) evaluate the effect of UHI mitigation strategies on outdoor thermal comfort in Baghdad, Iraq. In this study, three residential buildings selected which were Attached, Detached, and low-rise buildings. The detached pattern contains wide spaces between the buildings in the form of streets, pedestrian walkways. The attached pattern contained small spaces in comparison to the detached pattern. Low-rise building is spread throughout the site in the form of a ring. The unique factor of selecting these configurations was urbanized area, layout, and building characteristics. residential units have been selected.

A total of 18 scenarios include parameters such as green spaces, cool pavement, tree locations, and tree heights. The parametric simulation has been done with the help of simulation software ENVI-met which is a 3D tool that forecasts the microclimatic changes within the urban environments. The results of these scenarios were compared based on their effect on human thermal comfort. PMV has been used as an outdoor thermal comfort index. The main conclusion from the study are as follows:

- Urban geometry and vegetation were the most effective strategies which helped reduce human thermal discomfort significantly. Additionally, using cool pavement led to a slight increase in thermal discomfort.
- The shade from the trees and buildings led to a decrease in the PMV value by 2 to 4. Surfaces having higher albedo led to an increase in MRT because of increased solar radiation during the daytime.
- MRT was considered as the most important parameter affecting human thermal comfort as the MRT difference reached between 19°C and 26°C at the peak hour in most of the cases. The influence of other parameters (air temperature, relative humidity, wind speed) was not significantly large.
- The thermal comfort in outdoor spaces could not be achieved under extremely hot & dry conditions for daytime. However, the thermal discomfort decreased considerably by providing shading from trees and buildings.

This study provides a better understanding of the role of UHI mitigation strategies on human thermal comfort in the outdoor spaces and can help architects, building planners, and policymakers to generate guidelines of urban design in hot & dry climates.

5.3. Case Study-3: A Urban Canyon in Warm and Humid Climate of India

A study (De & Mukherjee, 2018) investigate the impact of orientations and canyon aspect ratios to improve the microclimate in the warm humid climate of Kolkata, India. Two neighborhoods had been identified as the study area with urban canyons running in East-West and North-South directions perpendicular to each other. The simulation model has been setup in ENVI-met and results are compared for different scenarios in terms of air temperature, MRT, wind speed, and PET. Only 2 parameters, orientation, and canyon aspect ratio have been considered. 4 orientations i.e., 0°, 15°, 30°, and 45° were considered, and a total number of 4 aspect ratios (H/W ratio) i.e., 0.6, 1.5, 2.0, 2.5 have been selected for parametric analysis. The outcomes of the study are as follows:

- Orientation and Canyon aspect ratio have a great impact on the urban microclimate that directly affects street-level thermal comfort
- A more open design with greater building height can substantially decrease the ambient air temperature and reduce the thermal stress due to better shading and high wind speed.
- A temperature reduction up to 5°C can be observed through optimization of the orientation and canyon aspect ratio in 30% of the area in question, 10-15°C in 15% of the area, and 15-20°C in 6% of the area.

It is also stated from the study that orientation and canyon aspect ratio can reduce the thermal discomfort, but it is difficult to bring the thermal comfort down within the comfort zone during peak time. It has been observed from the study that the outdoor thermal comfort variables are highly dependent on wind speed and direction of the flow but, as the wind flow pattern varies in the warm and humid climate zone, no generalized recommendations for orientation has been made.

5.4. Case Study-4: A Pedestrian Walkway in Cold Climate of Canada

This study (Taleghani & Berardi, 2018) focuses on the microclimate of one of the most important spaces in Toronto. The simulation has been performed for the hottest day as per the weather file. The pavement characteristics in open space have been investigated. There is a square which is covered by black pavement, and a small water pond exists in there. Three UHI mitigation scenarios were used to alter the land cover which includes increasing the albedo. The main inferences are explained below:

- When the albedo increases from 0.1 to 0.5, the air temperature is reduced up to 0.67°C due to the reflectivity of the material.
- Using water bodies, the air temperature gets reduced over the top of the water body but did not affect the thermal comfort situation in the square. The water pond reduced PET by up to 3.6°C.

The study shows that the increase of reflectivity reduced the air temperature, thus reducing the thermal comfort at the pedestrian level. Further, the findings showed that water bodies or water spray could be a promising strategy to mitigate UHI in places that are not dealing with water shortages.

6. Recommendations

6.1. Designing for Thermal Comfort

The effect of the thermal environment on different individuals might be varied, making the thermal comfort analysis of many users a difficult task. The primary factors determining thermal comforts such as air temperature, airspeed, relative humidity, and radiant temperatures are shared by both indoor and outdoor space, but they are subject to more extreme, intricate, and asymmetrical relational changes in outdoor conditions. These factors vary as per the geographical location, climatic type, and localized physical factors. Further challenges of determining thermal comfort levels in outdoor space include a substantial change in personal opinions and preferences for outside temperatures, which results in a considerably wider range of thermal comfort suitability. These preferences again depend upon geographical location and adaptation of people to certain situations (Givoni, et al., 2003; Johansson, et al., 2014; Brown, et al., 2015).

Many of these characteristics are out of planners' control when it comes to influencing thermal comfort in outdoor situations. However, designers must regard exterior space as available to a few users and have a restricted ability to monitor specific subjects at this degree of precision. (Chen & Ng, 2012; Walls, et al., 2015). Furthermore, the impact of design on many climatic factors is difficult to predict. For example, very large-scale initiatives such as national parks can alter ambient air temperature and relative humidity. However, in most design situations, the scale necessary to affect these parameters is impossible, especially within the constraints of existing urban site characteristics (Brown, et al., 2015).

Understanding the linkages of thermal comfort parameters in diverse climatic situations; limited control of these factors; and highly varied responses of users are among issues addressed by design that seeks to impact thermal comfort in exterior space. Another design problem is the creation of new spatial arrangements, which necessitates the prediction of future change as well as the measurement of current site circumstances. It is somewhat possible that the thermal comfort level may not be met in many situations. Thus, designers need a suitable index to use as an analytical tool that will indicate relative differences in the performance of the site concerning variable conditions (Givoni, et al., 2003; Chen & Ng, 2012; Walls, et al., 2015). The spectrum of commonly used thermal comfort indices provides numerous options for influencing the design, but they differ in how they handle location, size, existing conditions, and resource needs (Tan, et al., 2015). There are presently no suggestions for making appropriate thermal comfort indices for situations or advice on integrating these into the design, therefore, the use of indices as analytical design tools continues a space for additional research.

6.2. Rethinking the Thermal Comfort Standards

At present, the thermal comfort standards such as ASHRAE 55 (ASHRAE, 2010), ISO 7730, and EN 15251 (CEN, 2007), mostly depend on the physical, physiological, and adaptive methods. This dependency has hampered the development of logical associations between diverse thermal satisfaction factors (Bueno, et al., 2021), and due to these constraints, it has already been suggested that the traditional idea of thermal comfort that underpins thermal comfort standards should be reconsidered (Mamani, et al., 2022).

During the assessment of comfort data, this revision should look for methods to incorporate other understudied elements that may influence thermal satisfaction. As additional studies (Bueno, et al., 2021; Mamani, et al., 2022; Eslamirad, et al., 2022) include theories into the study design, analysis, and interpretation processes, new meanings are expected to arise and be incorporated into thermal comfort requirements. This will lead to a more in-depth understanding of comfort circumstances that are more indicative of thermal satisfaction in different situations. It raises questions as well about the necessity for defining local thermal comfort standards that are specific to outdoor circumstances and take into consideration contextual factors (Heng & Chow, 2019).

6.3. Impact of Climate on the Cooling Effect

Climate is one of the important factors for cooling energy. It has been noticed (Blazejczyk, et al., 2012; Cheung & Jim, 2018) that a greater cooling impact was attained when UTCI reducing strategies have been used in warmer climate conditions since these climates provide better potential for heat reduction (Aram, et al., 2020). A comprehensive study (Alexandri & Jones, 2008) explained that the green wall or green roof could reduce the urban air temperature for hot & dry climates but not very effective for warm and humid climates. (Wang, et al., 2015) indicated that the cooling impact of trees on hot & clear days was higher (almost double) than on cold & cloudy days. Likewise, the impact of high albedo in decreasing surface temperature was larger on bright days than overcast days as shown by observations from an experiment (Rosso, et al., 2016). Meanwhile, on sunny days, water has a greater cooling impact than on overcast days since the increased solar radiation on bright days offers more energy for water vaporization and a drop in air temperature (Chatzidimitriou & Yannas, 2015; Robitu, et al., 2006). Besides the ambient air temperature and solar radiation, the wind situation may affect the cooling impact via modifying heat transfer (Kalinowska, 2019). For instance, in a neighboring urban park having a normal wind velocity of 1.4 m/s, the reduction in surface temperature of reflective pavement can be as high as 7.2°C (Santamouris, et al., 2012).

6.4. Recommendations to Reduce Urban Heat Island

The urban heat mitigation strategies include the use of shading, construction of green infrastructures, modification of urban geometry, use of water bodies, and enhancement of albedo.

present the recommendations to mitigate the urban heat island determined from the experiments and research discussed in this chapter. The study recognizes supporting actions for each suggestion which are categorized as per indicated timeframes - short term: 1-3 years; and long term: 3-8 years.

Table 5. Holistic recommendations to reduce urban heat island for different timeframe

	Short term	**Long term**
Cool Roof	• Effective promotion and implementation of cool roofs by making it mandatory in rating standards rather than as a prescriptive measure. • Large and densely populated cities can take the initiative and establish programs on actions like cool roofs and pavements that would bring about a positive change in UHI.	• The increase in the roof albedo could reduce the absorption of solar radiation along with roof temperature and thus heat ingress into the buildings. • Setup a cool roof rating council that oversees a labelling program for roofing products and rates the reflectivity and emissivity for paint or coating.
Reducing off-street parking	Replacing surface parking lots with naturally vegetated pavements that transmit heat to the atmosphere.	Providing on-street parking, compact planning, and pedestrian-oriented development will promote transportation choices and will help in minimizing the size and number of parking lots
Promoting infill and higher density development	Growth within existing communities can protect open space and help counterbalance heat islands and their aftereffects	
Increasing public education and outreach	• UHI mitigation strategies should reflect local variation in the built environment, as well as local preferences and attitudes. • Policies should be customized to meet the requirements, based on stakeholder feedback, and efficiently conveyed to the public.	• A committee/organization could be formed to propagate the information to the community, seek advice, and incorporate problems and concerns into action plans • Committees formed to address UHI mitigation should include representatives from citizen groups, local government, non-governmental organizations, universities, and others concerned about the future growth of the community.

Conclusion

It has been recognized that outdoor thermal comfort is the growing field of concern for investigators because of the rapid growth, climate change, and population increase in urban areas. The main objective in climate-responsive urban design is to attain outdoor comfort. It is established that urbanization tends to exaggerate the adverse impacts of climate during the summertime because of the enhanced thermal capability, and shortage of water for evapotranspiration. This usually triggers thermal discomfort in Mediterranean cities during summertime. The use of thermal comfort indices in exterior space design can give valuable insight into the interactions between the fundamental variables that influence thermal feeling, but this can only be considered an opening move as the thermally comfortable urban spaces deals with a compilation of a variety of concerns.

Various mitigation measures have been adopted to decrease thermal stress associated with UHI, such as urban pattern design, vegetation plantation, highly reflective pavements, and cool roofs. The aforementioned strategies can reduce ambient air temperature, improve air quality, energy savings, increase greater thermal comfort, and add aesthetic advantages. The following inferences can be drawn from the study to diminish the UHI effect:

- Urban geometry alters the radiative and convective heat transfer in outdoor areas, or it can be said that small spaces offer better thermal comfort than open spaces in hot regions. Although the wind is low in dense areas, the effect of blocked radiation on the thermal environment surpasses the effect of decreased winds.
- North-South oriented streets propose an extended shading cycle than the East-West streets and roads with similar orientation to the wind direction are favorable in increasing wind speed and help achieve better outdoor thermal comfort.
- Vegetation scatters and absorbs sort wave radiation and reduces the wind and air temperature using transpiration and evaporation. Longwave radiation also can be decreased by using lawns/green walls.
- Reflective surfaces absorb less solar radiation and have a reduced surface temperature. The cool surface produces less long-wave radiation and lowers the air temperature. Though, some studies concluded that an increase in reflected surface balances the reduction

in long wave reduction, resulting in a harsh outdoor thermal comfort at pedestrian level.

- Evaporation from water bodies lowers the air temperature in outdoor locations, however, this is dependent on several parameters including size, geometry, position, wind direction, and the amount of the surroundings.

In summary, the urban geometry has a significant impact on the thermal environment in summer afterward vegetation and water bodies. The cool surface deteriorates the outdoor thermal comfort because of the greater exposed solar radiation. The study also supports that these strategies had a better cooling effect in hot environments.

Over the last century, more than 165 thermal comfort indices were established to assess thermal comfort based on different climatic zones. Many of the indices, however, are not used nowadays. Researchers must find out the correctness, usefulness, and pertinence as per the specific solution of the problem. Designers need to focus on the challenges of thermal performance in outdoor spaces to ensure these types of locations remain valuable in the future. Many current and growing tools and methodologies, including known thermal comfort indices, are available to deal with the variables that influence thermal comfort. Further research can be done on how these tools are best integrated into design processes. Researchers analyzed outdoor environments for a variety of reasons, including gaining a better knowledge of thermal comfort requirements, determining the influence of vegetation on thermal comfort requirements, evaluating the pertinence of indices, and exploring heat mitigation measures. The use of heat mitigation measures and plants has been demonstrated to increase thermal comfort. There is a lot of room for legislators and other stakeholders to use the findings of OTC research to promote urban open spaces. The use of heat mitigation measures provides policymakers, urban planners, government agencies, scientists, and horticulturists with a wide range of options for developing sustainable cities.

References

Ahmed, T., Kumar, P. & Mottet, L., 2021. Natural ventilation in warm climates: The challenges of thermal comfort, heatwave resilience and indoor air quality. *Renewable and Sustainable Energy Reviews,* Volume 138.

Akbari, H. & Konopacki, S., 2005. Calculating energy-saving potentials of heat-island reduction strategies. *Energy Policy,* 33(6), pp. 721-756.

Akbari, H. & Matthews, H. D., 2012. Global cooling updates: Reflective roofs and pavements. *Energy and Buildings,* Volume 55, pp. 2-6.

Aktas, Y. D., Wang, K., Zhou, Y., Othman, M., Stocker, J., Jackson, M., Hood, Ch., Carruthers, D., Talib Latif, M., Dina D'Ayala, D. and Hunt, J., 2020. Outdoor thermal comfort and building energy use potential in different land-use areas in tropical cities: Case of Kuala Lumpur. *Atmosphere,* 11(6), pp. 1-17.

Albatayneh, A., Alterman, D., Page, A. & Moghtaderi, B., 2018. The Impact of the Thermal Comfort Models on the Prediction of Building Energy Consumption. *Sustainability,* Volume 10.

Alexandri, E. & Jones, P., 2008. Temperature decreases in an urban canyon due to green walls and green roofs in diverse climates. *Building and Environment,* 43(4), pp. 480-493.

Algeciras, J., Rodríguez Algeciras, J. A., Coch, H., De la Paz Pérez, G., Chaos Yeras, M. Ch., Matzarakis, A., 2016. Human thermal comfort conditions and urban planning in hot-humid climates—The case of Cuba. *International Journal of Biometeorology,* 60(8), pp. 1151-64.

Ali, S. B. & Patnaik, S., 2018. Thermal comfort in urban open spaces: Objective assessment and subjective perception study in tropical city of Bhopal, India. *Urban Climate,* Volume 24, pp. 954-967.

Ali-Toudert, F. & Mayer, H., 2006. Numerical study on the effects of aspect ratio and orientation of an urban street canyon on outdoor thermal comfort in hot and dry climate. *Building and Environment,* 41(2), pp. 94-108.

Allegrini, J., Dorer, V. & Carmeliet, J., 2012. Influence of the urban microclimate in street canyons on the energy demand for space cooling and heating of buildings. *Energy and Buildings,* Volume 55, pp. 823-832.

Alonso, M. S., Fidalgo, M. R. & Labajo, J. L., 2007. The urban heat island in Salamanca (Spain) and its relationship to meteorological parameters. *Climate Research,* 34(1), pp. 39-46.

Andreou, E., 2013. Thermal comfort in outdoor spaces and urban canyon microclimate. *Renewable Energy,* pp. 182-188.

Aram, F., Solgi, E., Garcia, E. & Mosavi, A., 2020. Urban heat resilience at the time of global warming: evaluating the impact of the urban parks on outdoor thermal comfort. *Environmental Sciences Europe,* Volume 32.

Asadi, S., Fakhari, M. & Sendi, M., 2016. A study on the thermal behavior of traditional residential buildings: Rasoulian house case study. *Journal of Building Engineering,* Volume 7, pp. 334-342.

ASHRAE, 2010. *ASHRAE 55:2010 Thermal Environmental Conditions for Human Occupancy,* Atlanta: ASHRAE.

Ballinas, M. & Barradas, V. L., 2016. The Urban Tree as a Tool to Mitigate the Urban Heat Island in Mexico City: A Simple Phenomenological Model. *Journal of Environmental Quality,* 45(1), pp. 157-166.

BEE, 2017. *Energy Conservation Building Code,* New Delhi: Bureau of Energy Efficiency, Ministry of Power.

Berardi, U. & Wang, Y., 2016. The effect of a denser city over the urban microclimate: the case of Toronto. *Sustainability,* 8(8).

Binarti, F., Binarti, F., Donny Koerniawan, M., Triyadi, S., Sesotya Utami, S., Matzarakis, A., 2020. A review of outdoor thermal comfort indices and neutral ranges for hot-humid regions. *Urban Climate,* Volume 31.

Blazejczyk, K. Yoram Epstein, Y., Gerd Jendritzky, G., Staiger, H. and Tinz, B., 2012. Comparison of UTCI to selected thermal indices. *International Journal of Biometeorology,* Volume 56, pp. 515-535.

Blazejczyk, K., Jendritzky, G., Bröde, P., Fiala, D., Havenith, G., Epstein, Y., Psikuta, A. and Kampmann B., 2013. An introduction to the Universal Thermal Climate Index (UTCI). *Geographia Polonica,* 86(1), pp. 5-10.

Bowler, D. E., Buyung-Ali, L., Knight, T. M. & Pullin, A. S., 2010. Urban greening to cool towns and cities: A systematic review of the empirical evidence. *Landscape and Urban Planning,* 97(3), pp. 147-155.

Bröde, P., Błazejczyk, K., Fiala, D., Havenith, G., Holmér, I., Jendritzky, G., Kuklane, K. and Kampmann, B., 2013. The Universal Thermal Climate Index UTCI Compared to Ergonomics Standards for Assessing the Thermal Environment. *Industrial Health,* Volume 51, pp. 16-24.

Bröde, P. Fiala, D., Epstein, Y., Blazejczyk, K., 2009. *Calculating UTCI Equivalent Temperature.* Natick, MA, s.n.

Brown, R., Vanos, J., Kenny, N. & Lenzholzer, S., 2015. Designing urban parks that ameliorate the effects of climate change. *Landscape and Urban Planning,* Volume 138, pp. 118-131.

Bueno, A., Xavier, A. & Broday, E., 2021. Evaluating the Connection between Thermal Comfort and Productivity in Buildings: A Systematic Literature Review. *Buildings,* 11(6), p. 244.

Canan, F., Golasi, I., Ciancio, V., Coppi, M., Salatab, F., 2019. Outdoor thermal comfort conditions during summer in a cold semi-arid climate. A transversal field survey in Central Anatolia (Turkey). *Building and Environment,* Volume 148, pp. 212-224.

CEN, 2007. *EN 15251, Indoor environmental input parameters for design and assessment of energy performance of buildings addressing indoor air quality, thermal environment, lighting and acoustics,* Brussels, Belgium: European Committee for Standardization.

Charalampopoulos, I., Tsiros, I., Chronopoulou-Sereli, A. & Matzarakis, A., 2013. Analysis of thermal bioclimate in various urban configurations in Athens, Greece. *Urban Ecosystem,* 16(2), pp. 217-233.

Chatzidimitriou, A. & Yannas, S., 2015. Microclimate development in open urban spaces: the influence of form and materials. *Energy and Buildings,* Volume 108, pp. 156-174.

Chatzidimitriou, A. & Yannas, S., 2017. Street canyon design and improvement potential for urban open spaces; the influence of canyon aspect ratio and orientation on microclimate and outdoor comfort. *Sustainable Cities and Society,* Volume 33, pp. 85-101.

Chen, G., Rong, L. & Zhang, G., 2021. Unsteady-state CFD simulations on the impacts of urban geometry on outdoor thermal comfort within idealized building arrays. *Sustainable Cities and Society,* Volume 74.

Cheng, V., Ng, E., Chan, C. & Givoni, B., 2012. Outdoor thermal comfort study in a sub-tropical climate: a longitudinal study based in Hong Kong. *International Journal of Biometeorology,* Volume 56, pp. 43-56.

Chen, L. & Ng, E., 2012. Outdoor thermal comfort and outdoor activities: A review of research in the past decade. *Cities,* 29(2), pp. 118-125.

Chen, Y. & Matzarakis, A., 2018. Modified physiologically equivalent temperature—basics and applications for western European climate. *Theoretical and Applied Climatology,* Volume 132, pp. 1275-1289.

Cheung, P. & Jim, C., 2018. Comparing the cooling effects of a tree and a concrete shelter using PET and UTCI. *Building and Environment,* Volume 30, pp. 49-61.

Coccolo, S., Kampf, J., Scartezzini, J.-L. & Pearlmutter, D., 2016. Outdoor human comfort and thermal stress: A comprehensive review on models and standards. *Urban Climate,* Volume 18, pp. 33-57.

Coutts, A., Beringer, J. & Tapper, N., 2008. Investigating the climatic impact of urban planning strategies through the use of regional climate modelling: A case study for Melbourne, Australia. *International Journal of Climatology,* 28(14), pp. 1943-1957.

Crawford, R. & Stephen, A., 2015. *Designing with thermal comfort indices in outdoor sites.* Meloume, The Architectural Science Association and The University of Melbourne, pp. 1117-1128.

Davis, R., Knight, D., Hondula, D. & Knappenberger, P., 2006. *A comparision of biometeorological comfort indices and human mortality during heat waves in the united states.* San Diego, CA, s.n.

De Dear, R. & Pickup, J., 2000. An outdoor thermal comfort index (Out_SET*)–Part 1–The model and its assumptions. In: J. K. T. O. A. A. RJ de Dear, ed. *Biometeorology and Urban Climatology at the Turn of the Millennium.* Geneva: WMO.

de Freitas, C. R. & Grigorieva, E. A., 2021. A comparison and appraisal of a comprehensive range of human thermal climate indices. *International Journal of Biometeorology,* 61(3), pp. 487-512.

De, B. & Mukherjee, M., 2018. Optimisation of canyon orientation and aspect ratio in warm-humid climate: Case of Rajarhat Newtown, India. *Urban Climate,* Volume 24, pp. 887-920.

Dincer, I. & Rosen, M. A., 2013. Chapter 1 - Thermodynamic Fundamentals. In: *Exergy (Second Edition).* s.l.:Elsevier, pp. 1-20.

Din, M. F. M., Mohanadoss, P., Lee Y., Ossen D. R., 2014. Thermal comfort of various building layouts with a proposed discomfort index range for tropical climate. *Journal of Thermal Biology,* Volume 41, pp. 6-15.

Djamila, H., 2017. Indoor thermal comfort predictions: Selected issues and trends. *Renewable and Sustainable Energy Reviews,* Volume 74, pp. 569-580.

Djongyang, N., Tchinda, R. & Njomo, D., 2010. Thermal comfort: A review paper. *Renewable and Sustainable Energy Reviews,* 14(9), pp. 2626-2640.

Du, X., Bokel, R. & Debbelsteen, A., 2019. Spatial configuration, building microclimate and thermal comfort: A modern house case. *Energy and Buildings,* Volume 193, pp. 185-200.

Du, X., Bokel, R. & Dobbelsteen, A., 2014. Building microclimate and summer thermal comfort in free-running buildings with diverse spaces: A Chinese vernacular house case. *Building and Environment,* Volume 82, pp. 215-227.

Elnabawi, M. & Hamza, N., 2020. Outdoor Thermal Comfort: Coupling Microclimatic Parameters with Subjective Thermal Assessment to Design Urban Performative Spaces. *Buildings,* 10(12).

Emmanuel, R., Rosenlund, H. & Johansson, E., 2007. Urban shading—a design option for the tropics? A study in Colombo, Sri Lanka. *International journal of climatology.*

Epstein, Y. & Moran, D. S., 2006. Thermal comfort and the heat stress indices. *Industrial heatlh,* 44(3), pp. 388-398.

Eslamirad, N., Sepúlveda, A., De Luca, F. & Sakari Lylykangas, K., 2022. Evaluating Outdoor Thermal Comfort Using a Mixed-Method to Improve the Environmental Quality of a University Campus. *Energies,* Volume 15.

Falasca, S. Virgilio Ciancio, V., Salata, F., Iacopo Golasi, I., Rosso, F., Curci, G., 2019. High albedo materials to counteract heat waves in cities: An assessment of meteorology, buildings energy needs and pedestrian thermal comfort. *Building and Environment,* Volume 163, p. 106242.

Fanger, P. O., 1972. *Thermal Comfort: Analysis and Applications in Environmental Engineering.* s.l.:Danish Technical Press, 1970.

Fayad, F., Maref, W. & Awad, M., 2021. Review of White Roofing Materials and Emerging Economies with Focus on Energy Performance Cost-Benefit, Maintenance and Consumer Indifference. *Sustainability,* Volume 13.

Fiala, D., Havenith, G., Bröde, P., Kampmann, B., Jendritzky, G., 2012. UTCI-Fiala multi-node model of human heat transfer and temperature regulation. *International Journal of Biometeorology,* 56(3), pp. 429-441.

Fiala, D., Lomas, K. & Stohrer, M., 2001. Computer prediction of human themoregulatory and temperature responses to a wide range of environmental conditions. *International journal of bio-meteorology,* 45(3), pp. 143-159.

Fong, C. S. Aghamohammadi, N., Ramakreshnan, L., Nik, L., Sulaiman, M., Mohammadi, P., 2019. Holistic recommendations for future outdoor thermal comfort assessment in tropical Southeast Asia: A critical appraisal. *Sustainable Cities and Society,* Volume 46.

Gagge, A., Fobelets, A. & Berglund, L., 1986. A standard predictive index of human response to the thermal environment. *ASHRAE Transactions,* Volume 92, pp. 709-731.

Gagge, A., Stolwijk, J. & Nishi, Y., 1977. An effective temperature scale based on a simple model of human physiological regulatory response. *ASHRAE Transactions,* Volume 77, pp. 247-257.

Garg, V., Kotharkar, R., Sathaye, J., Rallapalli, H., Kulkarni, N., Niranjan Reddy, Raoa, P., Sarkar, A., 2016. Assessment of the impact of cool roofs in rural buildings in India. *Energy and Buildings,* Volume 114, pp. 156-163.

Ghalhari, G. F., Dehghan, S. F., Shakeri, F., Abbasinia, M., Asghari, M., 2019. Thermal comfort and heat stress indices for outdoor occupations over 15 years: A case study from Iran. *Weather,* pp. 1-6.

Girgis, N., Elariane, S., Elrazik, M. A. & Abd Elrazik, M., 2016. Evaluation of heat exhausts impacts on pedestrian thermal comfort. *Sustainable Cities and Society,* Volume 27, pp. 152-159.

Givoni, B., Noguchi, M., Saaroni, H., Pochter, O., Yaacov, Y., Feller, N., Becker, S., 2003. Outdoor comfort research issues. *Energy and Buildings,* 35(1), pp. 77-86.

Golasi, I., Salata, F., Vollaro, E. d. L. & Coppi, M., 2018. Complying with the demand of standardization in outdoor thermal comfort: A first approach to the Global Outdoor Comfort Index (GOCI). *Building and Environment,* Volume 130, pp. 104-119.

Hart, M. A. & Sailor, D. J., 2009. Quantifying the influence of land-use and surface characteristics on spatial variability in the urban heat island. *Theoretical and Applied Climatology,* 95(4), pp. 397-406.

Heng, S. L. & Chow, W. T., 2019. How 'hot' is too hot? Evaluating acceptable outdoor thermal comfort ranges in an equatorial urban park. *International Journal of Biometeorology,* Volume 63, pp. 801-816.

Hinkle, K., 2016. *The Albedo Effect, Urban Heat Islands, and Cooling Down Your Playground.* s.l.:s.n.

Honjo, T., 2009. Thermal comfort in outdoor environment. *Global Environmental Research,* Volume 13, pp. 43-47.

Hoppe, P., 1993. Heat balance modelling. *Experientia,* 49(9), pp. 741-746.

Hoppe, P., 1999. The physiological equivalent temperature – a universal index for the biometeorological assessment of the thermal environment. *International Journal of Biometerology,* 43(2), pp. 71-75.

Hoppe, P., 2002. Different aspects of assessing indoor and outdoor thermal comfort. *Energy and Buildings,* 34(6), pp. 661-665.

Horrison, E. & Amirtham, L., 2016. Role of built environment on factors affecting outdoor thermal comfort–a case of T. Nagar, Chennai, India. *Indian Journal of Science and Technology,* 9(5), pp. 1-4.

Hove, V. L. W. A. Jacobs, C. M. J., Heusinkveld, B. G., Elbers, J. A., van Driel, B. L., Holtslag A. A. M., 2015. Temporal and spatial variability of urban heat island and thermal comfort within the Rotterdam agglomeration. *Building and Environment,* Volume 83, pp. 91-103.

Humphreys, M., 1978. Outdoor temperatures and comfort indoors. *Batiment International, Building Research and Practice,* 6(2).

Huynh, C. & Eckert, R., 2012. Reducing Heat and Improving Thermal Comfort through Urban Design – A Case Study in Ho Chi Minh City. *International Journal of Environmental Science and Development,* 3(5), pp. 480-485.

Ignatius, M., Wong, N. H. & Jusuf, S. K., 2015. Urban microclimate analysis with consideration of local ambient temperature, external heat gain, urban ventilation, and outdoor thermal comfort in the tropics. *Sustainable Cities and Society,* Volume 19, pp. 121-135.

ISO, 2005. *ISO 7730 International Standard, Moderate thermal environments - Determination of the PMV and PPD indices and specification of the conditions for thermal comfort,* s.l.: ISO.

Jacklitsch, B. Williams, W. J., Musolin, K., Coca, A., Kim, J-H., Turner, N., 2016. *NIOSH criteria for a recommended standard: occupational exposure to heat and hot environments,* s.l.: National Institute for Occupational Safety and Health, DHHS .

Jamei, E., Ossen, D. R., Seyedmahmoudian M., Sandanayake, M., Stojcevski, A., Horand, B., 2020. Urban design parameters for heat mitigation in tropics. Renewable and Sustainable Energy Reviews. *Renewable and Sustainable Energy Reviews,* Volume 134.

Jamei, E. & Rajagopalan, P., 2017. Urban development and pedestrian thermal comfort in Melbourne. *Solar Energy,* Volume 144, pp. 681-698.

Jendritzky, G., de Dear, R. & Havenith, G., 2012. UTCI—Why another thermal index?. *International Journal of Biometeorology,* 56(3), pp. 421-428.

Jeong, M., Park, S. & Song, G., 2016. Comparison of human thermal responses between the urban forest area and the central building district in Seoul, Korea. *Urban Forestry & Urban Greening,* Volume 15, pp. 133-148.

Johansson, E., Thorsson, S., Emmanuel, R. & Kruger, E., 2014. Instruments and methods in outdoor thermal comfort studies – The need for standardization. *Urban Climate,* 10(2), pp. 346-366.

Kakon, A., Mishima, N. & Kojima, S., 2009. Simulation of the urban thermal comfort in a high density tropical city: Analysis of the proposed urban construction rules for Dhaka, Bangladesh. *Building Simulation,* Volume 2.

Kalinowska, M. B., 2019. Effect of water–air heat transfer on the spread of thermal pollution in rivers. *Acta Geophys,* Volume 67, pp. 597-619.

Kántor, N., Chen, L. & Gál, C., 2018. Human-biometeorological significance of shading in urban public spaces—summertime measurements in Pécs, Hungary. *Landscape and Urban Planning,* Volume 170, pp. 241-255.

Katić, K., Li, R. & Zeiler, W., 2016. Thermophysiological models and their applications: a review. *Building and Enviornment,* Volume 106, pp. 286-300.

Kenny, N. A., Warland, J. S., Brown, R. D. & Gillespie, T. G., 2009. Assessing the performance of the COMFA outdoor thermal comfort model on subjects performing physical activity. *International Jounral of Biometeorology.*

Khare, V., Vajpai, A. & Gupta, D., 2021. A big picture of urban heat island mitigation strategies and recommendation for India. *Urban climate,* Volume 37.

Kim, S. W. & Brown, R. D., 2021. Urban heat island (UHI) variations within a city boundary: A systematic literature review. *Renewable and Sustainable Energy Reviews,* Volume 148.

Koc, C., Osmond, P. & Peters, A., 2018. Evaluating the cooling effects of green infrastructure: A systematic review of methods, indicators and data sources. *Solar Energy,* Volume 166, pp. 486-508.

Kohler, M., Blond, N. & Clappier, A., 2016. A city scale degree-day method to assess building space heating energy demands in Strasbourg Eurometropolis. *Applied Energy,* Volume 184, pp. 40-54.

Krpo, A., Salamanca, F., Martilli, A. & Clappier, A., 2010. On the impact of anthropogenic heat fluxes on the urban boundary layer: A two-dimensional numerical study. *Boundary-Layer Meteorology,* 136(1), pp. 105-127.

Kumari, P., Garg, V., Kumar, R. & Kumar, K., 2021. Urban Climate Impact of urban heat island formation on energy consumption in Delhi. *Urban Climate,* Volume 36.

Kumari, P., Kapur, S., Garg, V. & Kumar, K., 2020. Effect of surface temperature on energy consumption in a calibrated building: A case study of Delhi. *Climate,* 8(6).

Kumar, P. & Sharma, A., 2020. Study on importance, procedure, and scope of outdoor thermal comfort –A review. *Sustainable Cities and Society,* Volume 61.

Labdaoui, K., Mazouz, S., Reiter, S. & Teller, J., 2021. Thermal perception in outdoor urban spaces under the Mediterranean climate of Annaba, Algeria. *Urban Climate,* Volume 39.

Lai, A., Maing, M. & Ng, E., 2017. Observational studies of mean radiant temperature across different outdoor spaces under shaded conditions in densely built environment. *Building and Environment,* Volume 114, pp. 397-409.

Lai, D., Guo, D., Hou, Y., Lin, Ch., Chen, Q., 2014. Studies of outdoor thermal comfort in northern China. *Building and Environment,* Volume 77, pp. 110-118.

Lai, D. Liu, W., Gan, T., Liu, K. and Chen, Q., 2019. A review of mitigating strategies to improve the thermal environment and thermal comfort in urban outdoor spaces. *Science of the total environment,* Volume 661, pp. 337-353.

Lau, K., Ren, C., Ho, J. & Ng, E., 2016. Numerical modelling of mean radiant temperature in high-density sub-tropical urban environment. *Energy and Buildings,* Volume 114, pp. 80-86.

Li, J. & Liu, N., 2020. The perception, optimization strategies and prospects of outdoor thermal comfort in China: A review. *Building and Environment,* Volume 170.

Li, K., Li, X. & Yao, K., 2022. Outdoor Thermal Environments of Main Types of Urban Areas during Summer: A Field Study in Wuhan, China. *Sustainability,* Volume 14.

Linden, v. d., 2002. Thermal indoor climate building performance characterized by human comfort response. *Energy and Buildings,* 34(7), p. 737–744.

Lin, T.-P., Matzarakis, A. & Hwang, R.-L., 2010. Shading effect on long-term outdoor thermal comfort. *Building and Environment,* 45(1), pp. 213-221.

Lin, Y., Wang, Z., Jim, Y. J., LiJ., Deng, J., Liu, J., 2020. Water as an urban heat sink: blue infrastructure alleviates urban heat island effect in mega-city agglomeration. *Journal of cleaner production,* Volume 262, p. 121411.

Mamani, T., Herrera, R., Rivera, F. & Atencio, E., 2022. Variables That Affect Thermal Comfort and Its Measuring Instruments: A Systematic Review. *Sustainability,* Volume 14.

Manteghi, G., Limit, H. B. & Remaz, D., 2015. Water bodies an urban microclimate: a review. *Mathematical Models and Methods in Applied Sciences,* Volume 9, pp. 1-12.

Matzarakis, A. & Amelung, B., 2008. Physiological Equivalent Temperature as Indicator for Impacts of Climate Change on Thermal Comfort of Humans. In: *Seasonal Forecasts, Climatic Change and Human Health.* s.l.:Springer Science+Business Media B.V., pp. 161-172.

Mayer, H., Dostal P. and Holst J., 2008. Human thermal comfort in summer within an urban street canyon in Central Europe. *Meteorologische Zeitschrift,* 17(3), pp. 241-250.

Mechartes, 2019. *Pedesterian thermal comfort - Best practices,* New Delhi: Mechartes.

Mijorski, S., Cammelli, S. & Green, J., 2019. A hybrid approach for the assessment of outdoor thermal comfort. *Journal of Building Engineering,* Volume 22, pp. 147-153.

Milošević, D., Bajšanski, I. & Savić, S., 2017. Influence of changing trees locations on thermal comfort on street parking lot and footways. *Urban Forestry & Urban Greening,* Volume 23, pp. 113-124.

Nasir, R. A., Ahmad, S. S. & Ahmed, A. Z., 2013a. Physical activity and human comfort correlation in an urban park in hot and humid conditions. *Procedia – Social and Behavioral Sciences,* Volume 105, pp. 598-609.

Nasir, R., Ahmad, S. & Ahmed, A. Z., 2013b. Perceived and measured adaptive thermal comfort at an outdoor shaded recreational area in Malaysia. *Advanced Materials Research,* pp. 1083-1086.

Niu, J., Liu, J., Lee, T., Lin, Z. J., Mak, Ch., Tse, K-T., Tang, B., Kwok, K. C. S., 2015. A new method to assess spatial variations of outdoor thermal comfort: Onsite monitoring results and implications for precinct planning. *Building and Environment,* Volume 91, pp. 263-270.

O'Brien, W. & Gunay, H. B., 2014. The contextual factors contributing to occupants' adaptive comfort behaviors in offices – A review and proposed modeling framework. *Building and Environment,* Volume 77, pp. 77-87.

Oke, T., 1981. Canyon geometry and nocturnal urban heat island: comparison of scale model and field observations. *Journal of Climatology,* Volume 1, pp. 237-254.

Oke, T., 1988. The urban energy balance. *Progress in Physical Geography,* 12(4), pp. 471-508.

Pandey, P., 2009. A study of the summertime urban heat island over Delhi. *International Journal of Sustainability Science and Studies,* 1(1), pp. 27-34.

Panferov, O., Knyazikhin, Yu., Myneni, R. B., Szarzynski, J., 2001. The Role of Canopy Structure in the Spectral Variation of Transmission and Absorption of Solar Radiation in Vegetation Canopies. *IEEE Transactions on Geoscience and Remote Sensing,* 39(2), pp. 241-253.

Pearlmutter, D., Berliner, P. & Shaviv, E., 2006. Physical modeling of pedestrian energy exchange within the urban canopy. *Building and Environment,* 41(6), pp. 783-795.

Pigliautile, I., Pisello, A. L. & Bou-Zeid, E., 2020. Humans in the city: Representing outdoor thermal comfort in urban canopy models. *Renewable and Sustainable Energy Reviews,* Volume 133.

Psikuta, A., Fiala, D., Laschewski, G. and Jendritzky, G., 2012. Validation of the Fiala multi-node thermophysiological model for UTCI application. *International Journal of Biometeorology,* Volume 56, pp. 443-460.

Qaid, A., Lamit, H., Ossen, D. & Shahminan, R., 2016. Urban heat island and thermal comfort conditions at micro-climate scale in a tropical planned city. *Energy and Buildings,* Volume 133, pp. 577-595.

Rijal, H. B., Tuohy, P., Humphreys, M. A., Nicol, J. F., Samuel, A. and Clarke, J. 2007. Using Results from Field Surveys to Predict the Effect of Open Windows on Thermal Comfort and Energy Use in Buildings. *Energy and Buildings,* Volume 39, pp. 823-836.

Robitu, M., Musy, M., Inard, C. & Groleau, D., 2006. Modeling the influence of vegetation and water pond on urban microclimate. *Solar Energy,* 80(4), pp. 435-447.

Rose, D., Wu, H. & Beyers, M., 2010. *Spatial and temporal computation of thermal comfort inputs in outdoor spaces.* s.l., s.n.

Rosso, F., Pisello, A., Cotana, F. & Ferrero, M., 2016. On the thermal and visual pedestrians' perception about cool natural stones for urban paving: A field survey in summer conditions. *Building and Environment,* Volume 107, pp. 198-214.

Saaroni, H. & Ziv, B., 2010. Estimating the urban heat island contribution to urban and rural air temperature differences over complex terrain: Application to an Arid city. *Journal of Applied Meteorology and Climatology,* 49(10), p. 2159–2166.

Salman, A. & Saleem, Y., 2021. The effect of Urban Heat Island mitigation strategies on outdoor human thermal comfort in the city of Baghdad. *Frontiers of Architectural Research,* 10(4), pp. 838-856.

Santamouris, M., 2014. Cooling the cities – A review of reflective and green roof mitigation technologies to fight heat island and improve comfort in urban environments. *Solar Energy,* Volume 103, pp. 682-703.

Santamouris, M., Gaitani, N., Spanou, A., Saliari, M., Giannopoulou, K., Vasilakopoulou, K., Kardomateas, T., 2012. Using cool paving materials to improve microclimate of urban areas–design realization and results of the Flisvos project. *Building and Environment,* Volume 53, pp. 128-136.

Schellen, L., Loomans, M. G. L. C., Kingma, B. R. M., de Wit, M. H., Frijns, A. J. H., van Marken Lichtenbelt, W. D., 2013. The use of a thermophysiological model in the built environment to predict thermal sensation: Coupling with the indoor environment and thermal sensation. *Building and Environment,* Volume 59, pp. 10-22.

Shickman, K. & Rogers, M., 2020. Capturing the true value of trees, cool roofs, and other urban heat island mitigation strategies for utilities. *Energy Efficiency,* 13(3), pp. 407-418.

Shih, W., Lin, T., Tan, N. & Liu, M., 2017. Long-term perceptions of outdoor thermal environments in an elementary school in a hot-humid climate. *International Journal of Biometeorology,* Volume 61, pp. 1657-1666.

Shimazaki, Y., Yoshida, A., Suzuki, Y., Kawabata, T., Imai, D., Kinoshita, S., 2011. Application of human thermal load into unsteady condition for improvement of outdoor thermal comfort. *Building and Environment,* 46(8), pp. 1716-1724.

Shooshtarian, S., Lam, C. K. C. & Kenawy, I., 2020. Outdoor thermal comfort assessment: A review on thermal comfort research in Australia. *Building and Environment,* Volume 177.

Spagnolo, J. & de Dear, R., 2003. A field study of thermal comfort in outdoor and semi-outdoor environments in subtropical Sydney Australia. *Building and Environment,* 38(5), pp. 721-738.

Spagnolo, J. & De dear, R., 2003. A human thermal climatology of subtropical Sydney. *International Jounral of Climatology,* 23(11), pp. 1385-1395.

Syafii, N. I., Ichinose, M., Hien Wong, N., Kumakura, E., Kardinal Jusuf S., Chigusa K., 2016. Experimental Study on the Influence of Urban Water Body on Thermal Environment at Outdoor Scale Model. *Procedia Engineering,* Volume 169, pp. 191-198.

Taha, H., Sailor, D. & Akbari, H., 1992. *High-albedo materials for reducing building cooling energy use,* United States: s.n.

Taleghani, M., 2018. Outdoor thermal comfort by different heat mitigation strategies- A review. *Renewable and Sustainable Energy Reviews,* 81(2), pp. 2011-2018.

Taleghani, M. & Berardi, U., 2018. The effect of pavement characteristics on pedestrians' thermal comfort in Toronto. *Urban Climate,* Volume 24, pp. 449-459.

Tan, C., Wong, N. & Jusuf, S., 2013. Outdoor mean radiant temperature estimation in the tropical urban environment. *Building and Environment,* Volume 127, pp. 52-64.

Tan, Z., Lau, K. K.-L. & Ng, E., 2015. Urban tree design approaches for mitigating daytime urban heat island effects in a high-density urban environment. *Energy and Buildings,* Volume 114, pp. 265-274.

United Nations, 2018. *68% of the world population projected to live in urban areas by 2050, says UN.* [Online] Available at: https://www.un.org/development/desa/en/news/population/2018-revision-of-world-urbanization-prospects.html [Accessed 10 03 2022].

Walls, W., Parker, N. & Walliss, J., 2015. *Designing with thermal comfort indices in outdoor sites.* s.l., The Architectural Science Association and The University of Melbourne, pp. 1117-1128.

Wang, Y. & Akbari, H., 2016. The effects of street tree planting on Urban Heat Island mitigation in Montreal. *Sustainable Cities and Society,* Volume 27, pp. 122-128.

Wang, Y., Bakker, F., de Groot, R. S., Wortche, H., Leemans, R., 2015. Effects of urban trees on local outdoor microclimate: synthesizing field measurements by numerical modelling. *Urban Ecosystems,* Volume 18, pp. 1305-1331.

Wang, Y., Berardi, U. & Akbari, H., 2015. The Urban Heat Island Effect in the City of Toronto. *Procedia Engineering,* Volume 118, pp. 137-144.

Wang, Y., Berardi, U. & Akbari, H., 2016. Comparing the effects of urban heat island mitigation strategies for Toronto, Canada. *Energy and Buildings,* Volume 114, pp. 2-19.

Watanabe, S., N. K., Ishii, J. & Horikoshi, T., 2014. Evaluation of outdoor thermal comfort in sunlight, building shade, and pergola shade during summer in a humid subtropical region. *Building and Environment,* Volume 82, pp. 556-565.

Wong, N., Chen, Y., Ong, C. & Sia, A., 2003. Investigation of thermal benefits of rooftop garden in the tropical environment. *Building and Environment,* 38(2), pp. 261-270.

Wong, N. H., Jusuf, S. K. & Tan, C. L., 2011. Integrated urban microclimate assessment method as a sustainable urban development and urban design tool. *Landscape and Urban Planning,* 100(4), pp. 386-389.

Xi, T., Li, Q., Mochida, A. & Meng, Q., 2012. Study on the outdoor thermal environment and thermal comfort around campus clusters in subtropical urban areas. *Building and Environment,* Volume 52, pp. 162-170.

Yang, J., Shi, B., Xia, G., Xue, Q., Cao, S-J., 2020. Impacts of Urban Form on Thermal Environment Near the Surface Region at Pedestrian Height: A Case Study Based on High-Density Built-Up Areas of Nanjing City in China. *Sustainability,* 12(5).

Yang, R. & Wang, L., 2013. Development of multi-agent system for building energy and comfort management based on occupant behaviors. *Energy and Buildings,* Volume 56, pp. 1-7.

Yang, W., Wong, N. H. & Jusuf, S. K., 2013. Thermal comfort in outdoor urban spaces in Singapore. *Building and Environment,* Volume 59, pp. 426-435.

Yang, W., Wong, N. & Li, C., 2016. Effect of Street Design on Outdoor Thermal Comfort in an Urban Street in Singapore. *Journal of Urban Planning and Development,* 142(1).

Yuan, C. & Ng, E., 2012. Building porosity for better urban ventilation in high-density cities–a computational parametric study. *Building and Environment,* Volume 50, pp. 176-189.

Zhang, A., Bokel, R., van den Dobbelsteen, A., Sun, Y., Huang, Q., Zhang, Q., 2017. An integrated school and schoolyard design method for summer thermal comfort and energy efficiency in Northern China. *Building and Environment,* Volume 124, pp. 369-387.

Zhu, Y., Fan, X., Wang, C. & Sang, G., 2018. Analysis of Heat Transfer and Thermal Environment in a Rural Residential Building for Addressing Energy Poverty. *Applied Sciences,* 8(11), p. 2077.

Ziaul, S. & Pal, S., 2019. Assessing outdoor thermal comfort of English Bazar Municipality and its surrounding, West Bengal, India. *Advances in Space Research,* 64(3), pp. 567-580.

Chapter 6

Investigating the Cooling Load of a University Building: A Case Study in Darwin

Mohit Saini[1], Anne W. M. Ng[1,2], Hing-Wah Chau[3,4], Muhammad Atiq Ur Rehman Tariq[3,4], Elmira Jamei[3,4], Saeed Kimiaei[5], Raphael Chukwuka Nnachi[6], Mostafizur Rahman[1] and Hooman Mehdizadeh-Rad[1,2,*]

[1]College of Engineering, Information Technology and Environment, Charles Darwin University, Brinkin, Australia
[2]Energy and Resources Institute, Charles Darwin University, Brinkin, Australia
[3]College of Engineering and Science, Victoria University, Melbourne, Australia
[4]Institute for Sustainable Industries & Liveable Cities, Victoria University, Melbourne, Australia
[5]Department of Mechanical Engineering, Sharif University of Technnology, Azadi Ave, Tehran, Iran
[6]Faculty of Biological Sciences, Alex Ekwueme Federal University, Ndufu Alike Ikwo, Abakaliki P.M.B. 1010, Nigeria

Abstract

The human body resembles a heat engine as it constantly generates heat. The thermal envelope of a human must be maintained at optimum levels for maximum performance and comfort. In hot environments, extra heat is required to be expelled from the body while in the colder environment, the heat must be retained. Contradiction to both these conditions leads to

[*]Corresponding Author's Email: hooman.mehdizadehrad@cdu.edu.au.

In: Urban Heat Islands Reexamined
Editors: Satya Prakash and Anne W. Ng
ISBN: 979-8-88697-215-3

discomfort. Thermal condition is thus defined as mind satisfaction, as assessed by the American Society of Heating, Refrigerating and Air-Conditioning Engineers (ASHRAE) standards, leading to human satisfaction at utmost comfort levels for temperatures between 22°C to 27°C and relative humidity between 40% and 60%.

This chapter investigates the reduction of cooling load of a university building in Darwin, Australia, as a case study. This will be achieved with the use of the Revit software and tested in EnergyPlus in accordance with the climate of Darwin to achieve the desired comfort levels and results. Simulations are made by varying the construction materials of the walls and ceilings, glazing the windows, cool paint coating, solar panel installation, and glass replacement of windows for the acquirement of necessitated results.

The results thus obtained will provide a deeper understanding of the thermal comfort and cooling load. Based on these results, conclusions will be drawn by suggesting various alternatives for the improvement of human comfort and reducing energy consumption simultaneously.

Keywords: thermal comfort, revit, energyplus, heat mitigation, indoor space, Darwin

List of Abbreviations

ANSI	American National Standards Institute
ASHRAE	American Society of Heating, Refrigerating and Air-Conditioning Engineers
CFM	cubic feet per minute
Clo	Clothing insulation
CLF	Cooling load factor
CLTD	Cooling load temperature differential
NIOSH	National Institute of Occupational Safety and Health
SR	Solar reflectance
TFM	Transfer function method

1. Introduction

Thermal comfort is essential for human life. In unfavourable thermal environments, the human body lacks productivity. Humans can feel tired if it is too hot and can be restless and distracted if it is too cold. Appropriate temperature and humidity combined with appropriate air velocity contribute to the thermal comfort of the human body. For artificial thermal comfort, air conditioning system is installed for achieving the desirable range of temperature and humidity within a confined space with the expenditure of electricity. The amount of heat that is required to be removed from the space is termed as cooling load. Now with the necessity to stay in the thermal comfort, air conditioning units are provided in almost every building for the sake of benefits of humans which leads to higher energy consumption.

Buildings consume approximately 40% of the energy produced around the globe and are responsible for 30% of carbon dioxide emission. Most of the energy is consumed in providing the thermal comfort to the occupants, especially due to the widespread use of the HVAC equipment. If this trend continues, the globe can face energy crisis soon. Example can be taken from that of China. During the period of 1978 till 2010 China's total energy requirement jumped from 570 to 3200 million tonnes of coal (5.6% annual growth). Although the energy use of China is not at par with US, yet it overtook USA and became the largest energy consumer and carbon dioxide emitter of the 2009 (Yang et al., 2014). It is estimated that by the year 2050, China's primary energy requirement would increase to 6200 million tonnes of coal with the emissions reaching as high as 10^{10} tonnes of carbon dioxide emission (Chai & Zhang, 2010). Furthermore, it was also estimated that energy consumption in the developing countries such as Middle east, South America, Africa will surpass to that of developed countries such as Japan, Australia and New Zealand (Pérez-Lombard et al., 2008).

Facing the pressing issue of the urban heat island effect, the Northern Territory Government funded the research activities leading to the publication of the Heat Mitigation Program Final Report in 2017 (Mattheos et al., 2017). Energy consumption for cooling and peak loads is one of the key factors that contributes to the urban heat island effect (City of Darwin, 2022). In this chapter, we will investigate the reduction of the cooling load for the Pink 7 Building of Charles Darwin University in Darwin as a case study.

1.1 Darwin Weather

Darwin is the capital city of Northern territory state of Australia. It experiences a tropical savanna climate with a distinct wet and dry climate. The dry season lasts from May to September with sunny days and minimalistic to no rainfall. The days are sunny and the average humidity is approximately 30%. May to

October have the lowest amount of Rainfall, ranging from 0 to 0.5 mm. July has the lowest mean rainfall (1.1mm). With the average annual temperature of 23.2°C, July is the coldest month in which the mean minimum temperature drops to 19.3°C. However, the suburbs away from the coast can sometimes log the temperature as low as 5°C (Meteorology, 2022).

The wet season is comprised of tropical cyclones and seasonal rains, with most of the rainfall occurring between December to March, thunderstorms are common during that period and the relative humidity reaches to about 70% during those periods. Although the rain is not seen every day, but clouds are often (Evans et al., 2014). The hottest month is November with average temperature ranging between 30 to 35°C. The high temperature and high humidity lead to a higher level of discomfort.

1.2. Thermal Comfort

Research in thermal comfort is important as it comprises of several areas of study such as building physics, physiology, mechanical engineering, and psychology. There are many reasons for understanding the importance of thermal comfort: providing satisfactory conditions for end users, improving indoor air quality, achieving energy saving, and improving work efficiency of workers (Sayigh & Marafia, 1998; Taleghani et al., 2013).

Thermal comfort is defined by the ANSI/ASHRAE Standard 55-1992 as the state of mind in agreement to the thermal environment (ASHRAE, 2012). The ASHRAE handbook provides the detailed information regarding the human thermal comfort. Comfort is defined as the standard specific condition in which 80 percent or more of the occupants would find the environment to be thermally acceptable, thereby considering it as a subjective matter, depending upon the judgement of those affected (McQuiston, 2004). The factors that stimulate the thermal balance of a person include dry bulb temperature, humidity, relative velocity, temperature in contact to the part of body, exchanging heat in the form of radiation

1.3. Cooling Load

The cooling load is defined as the rate of energy removal from the space to maintain the temperature and humidity at the required design values. Cooling load calculations are required to provide the data for equipment selection, evaluate possible load reduction through the output data, and analysis of partial load for system design, control and operation (Bhatia, 2004).

In the calculation for the cooling load, transient analysis is used. For better understanding of cooling load, both the heat gain and heat extraction rate are considered. Heat gain is mainly due to various factors: solar radiation through openings, heat conduction and convection, ventilation and infiltration air, sensible heat, latent heat gain, etc. (McQuiston, 2004)

For the calculations of the cooling load, the unsteady processes must be considered as the peak conditions occur during the daytime and external conditions vary throughout the day due to solar radiation. All internal sources are crucial for the calculation of the cooling load, neglecting them leads to under-estimation of the result and the required cooling capacity may not be achieved (Bhatia, 2004). Whenever the load is to be calculated for the cooling load estimation, numerous factors need to be considered, such as: dimensions of walls and roofs, infiltration rate, and equipment heat gains.

1.3.1. Cooling Load Components

The cooling load of the space comprises of the heat travelling through the building envelope (walls, roofs, floors, doors, windows, etc.) along with the heat generated by the equipment, occupants, and the lights. Heat transfer from the surroundings is referred as external load while the one generated by the occupants and the equipment is known as internal loads. When the cooling load of the building is considered, both sensible heat and latent heat should be taken into consideration. For the complete calculation of the cooling load, the following three methods are used (Bhatia, 2004):

- Transfer function method (TFM) – it requires the use of the spreadsheet or a computer software as it is one of the most complex methods for the calculation of the cooling load as proposed by ASHRAE.
- Cooling load temperature differential/cooling load factor (CLTD/CLF) – It uses tabulated data for the ease of calculations and is derived from the TFM.
- Total Equivalent temperature differential/ Time-Averaging – This is a simple method for the calculation purposes and was used prior to the CLTD/CLF method.

Climate data for Darwin Airport (1991–2020 Averages, Extremes 1941–present)													[hide]
Month	Jan	Feb	Mar	Apr	May	Jun	Jul	Aug	Sep	Oct	Nov	Dec	Year
Record high °C (°F)	36.1 (97.0)	36.0 (96.8)	36.0 (96.8)	36.7 (98.1)	36.0 (96.8)	34.6 (94.3)	34.8 (94.6)	37.0 (98.6)	37.7 (99.9)	38.9 (102.0)	37.3 (99.1)	37.0 (98.6)	38.9 (102.0)
Average high °C (°F)	31.9 (89.4)	31.7 (89.1)	32.2 (90.0)	33.0 (91.4)	32.3 (90.1)	31.1 (88.0)	31.1 (88.0)	31.8 (89.2)	33.0 (91.4)	33.8 (92.8)	33.7 (92.7)	32.9 (91.2)	32.4 (90.3)
Daily mean °C (°F)	28.4 (83.1)	28.3 (82.9)	28.5 (83.3)	28.5 (83.3)	27.3 (81.1)	25.5 (77.9)	25.2 (77.4)	25.8 (78.4)	27.9 (82.2)	29.3 (84.7)	29.5 (85.1)	29.2 (84.6)	27.8 (82.0)
Average low °C (°F)	25.0 (77.0)	25.0 (77.0)	24.8 (76.6)	24.1 (75.4)	22.3 (72.1)	20.0 (68.0)	19.3 (66.7)	19.8 (67.6)	22.8 (73.0)	24.8 (76.6)	25.3 (77.5)	25.5 (77.9)	23.2 (73.8)
Record low °C (°F)	20.2 (68.4)	17.2 (63.0)	19.2 (66.6)	16.0 (60.8)	13.8 (56.8)	12.1 (53.8)	10.4 (50.7)	13.0 (55.4)	14.3 (57.7)	19.0 (66.2)	19.3 (66.7)	19.8 (67.6)	10.4 (50.7)
Average rainfall mm (inches)	473.1 (18.63)	419.1 (16.50)	314.9 (12.40)	104.7 (4.12)	21.4 (0.84)	2.2 (0.09)	1.0 (0.04)	0.8 (0.03)	11.7 (0.46)	65.5 (2.58)	145.5 (5.73)	275.6 (10.85)	1,835.5 (72.27)
Average rainy days	22.3	21.1	19.6	10.0	2.4	0.6	0.2	0.5	2.3	7.7	13.2	17.8	117.7
Average afternoon relative humidity (%)	71	74	67	52	41	36	36	38	47	51	58	66	53
Mean monthly sunshine hours	176.7	162.4	213.9	264.0	300.7	303.0	319.3	325.5	297.0	294.5	255.0	198.4	3,110.4

Figure 1. Climate data for Darwin (Ref. Bureau of Meteorology).

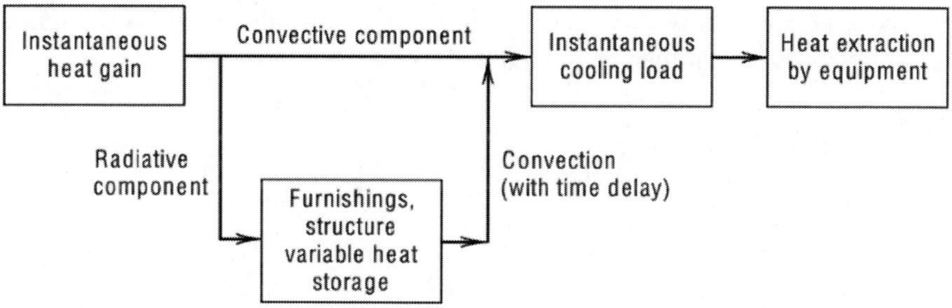

Figure 2. Relation of heat gain and cooling load.

For each of the method described above, several benefits and limitations are associated. Among which, simplicity and accuracy are contradictory objectives which are the main objectives to be fulfilled. Simple methods do not tend to be quite accurate. For strict cooling load calculations, the most practical method is the implementation of the CLTD/ CLF method, which generates the most conservative results as per the peak load of the equipment (Bhatia, 2004).

1.3.2. Load Design

To calculate the cooling load, detailed building information is required along with the location, weather data, internal design information, indoor conditions, working hours, occupants, and equipment details. The environmental factors that affect the thermal comfort in an air-conditioned space are: metallic rate, indoor air temperature (T_r), mean radiant temperature (T_{rad}), relative humidity of the indoor air which is represented in percentage, air velocity, and clothing insulation (clo) (Bhatia, 2004; Sait, 2013). A comfort indoor environment is recommended to be designed as per the ANSI/ASHRAE Standards 55.

The National Institute of Occupational Safety and Health (NIOSH) stated that the most common complaint about the quality of indoor air is the complaint of the adequate ventilated air, which can be improved by eliminating the source of air pollution, improving the efficiency of air filtration and increasing the ventilated air intake (Luksamijarulkul & Pipitsangjan, 2015).

Besides air quality, building pressurization should be taken into account. Most air conditioners are designed to maintain a slightly higher pressure in the space than the surroundings to reduce infiltration and prevent any untreated air from entering the space directly. For restrooms, workshops, laboratories, etc. a slightly negative pressure is maintained to hamper the diffusion of potential pollutants into surrounding areas (Bhatia, 2004).

1.4. Heat Mitigation Strategies

Heat mitigation strategies are deployed to remove excessive heat from a space to achieve thermal comfort for majority of occupants. There are various ways for heat mitigation (A. M. Omer, 2008)

1.4.1. Airflow Regulations

This is an energy efficient technique to monitor the thermal comfort and regulate the temperature of the space. A cooling sensation is noticeable when the airflow of 0.2 m/sec is experienced by the occupant (A. Omer, 2002). Natural ventilation from chimneys can induce passive airflow leading to cool sensation.

1.4.2. Wind Towers

They are a suitable option for summer comfort in hot arid regions. They can maintain natural ventilation in the building due to wind and buoyancy effects. The structure gets cooled by the radiative effect of the sky and internally with the cool air.

1.4.3. Solar Control

The best method to reduce the incoming heat gain by the solar radiation is by diverting the incident heat from the building using panels and shades.

1.4.4. Green Roof

It is the technique of planting trees on the roof of the building as to reduce the effects of solar heat gain through the roof. The vegetation provides the shade along with removal of heat from the surrounding by evapotranspiration and also aids in cooling the ambient air (Lepp, 2008).

1.4.5. Green Walls

Similar to the green roof, it is the technique of planting vegetation vertically on the walls that create a microclimate that lowers the temperature of the building envelope and improves the energy efficiency.

1.4.6. Cool/Reflective Roofs and Walls

It is the technique of deploying reflective material paint on the roof that can reflect the heat away from the space and reduce the cooling load (Rizwan et al., 2008). Reflective painting and coating are characterized by high solar reflectance (SR) and high infrared emittance. When a surface with high solar reflectance and infrared emittance is exposed to solar radiation, it has a lower temperature as compared to a similar wall with lower solar reflectance and infrared emittance values (Santamouris et al., 2011)

1.4.7. Energy Efficient Appliances

One of the simplest means to mitigate heat is using energy rated appliances by the EPA's Energy star label. Thigh energy star devices generate less heat as compared to the lower star and help to lower the internal gain by the space.

1.4.8. Building Construction

The most useful aspect that can reduce the cooling load of a building is the construction parameter. The opening equipped with shading and cool material help reduce the heat absorbed by the space. The space is less likely to absorb and radiate back the heat through the walls and roof if it has a higher reflectivity (albedo value) and emissivity (Prado & Ferreira, 2005).

1.4.9. Windows

They are the weakest point in a space's thermal insulation. The attributes of windows include the U-value, air leakage and transmittance. By changing the reflection of the window with the help of tinting or glazing or selective coating, the transmittance into interior space is reduced, leading to less direct solar heat gain (Nithiarasu et al., 2016). Thermal insulation of the windows can also be improved by the use of double or triple glazed windows to reduce absorptivity and increase insulation of windows.

2. Research Method

2.1. Building Modelling

Figure-3 shows the block level map of the Charles Darwin University. The Pink 7 Building was selected in this study for investigation (as encircled in the Figure 3).

The digital modelling of the Pink 7 Building was produced using Revit software. The pink 7 building is located in Casuarina campus, which is the University's largest campus and where most of the institutional activities takes place (Mamtora, 2011). The building contains classrooms and testing facility, so it provides enough data such as occupants, more lightings and various equipment for the calculation of the building loads.

Based on the weather of Darwin, the model clearly depicts the sun path sun throughout the day, shadow projection and the overall configuration of the building (Figure 4). After digital modelling, the building load factors, such as occupants, lighting and the equipment, were then calculated.

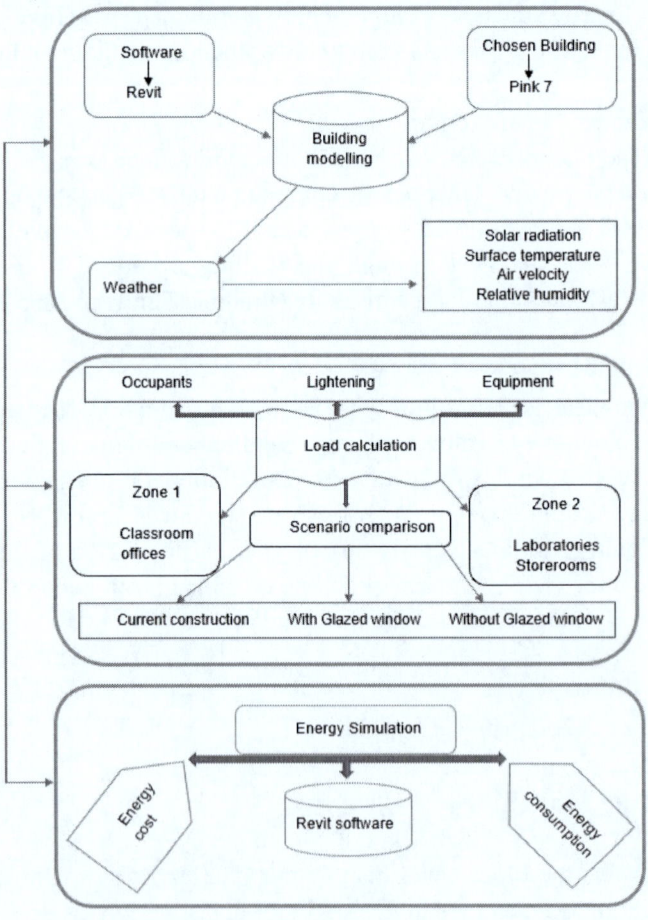

Figure 3. Flowchart of the adopted study approach.

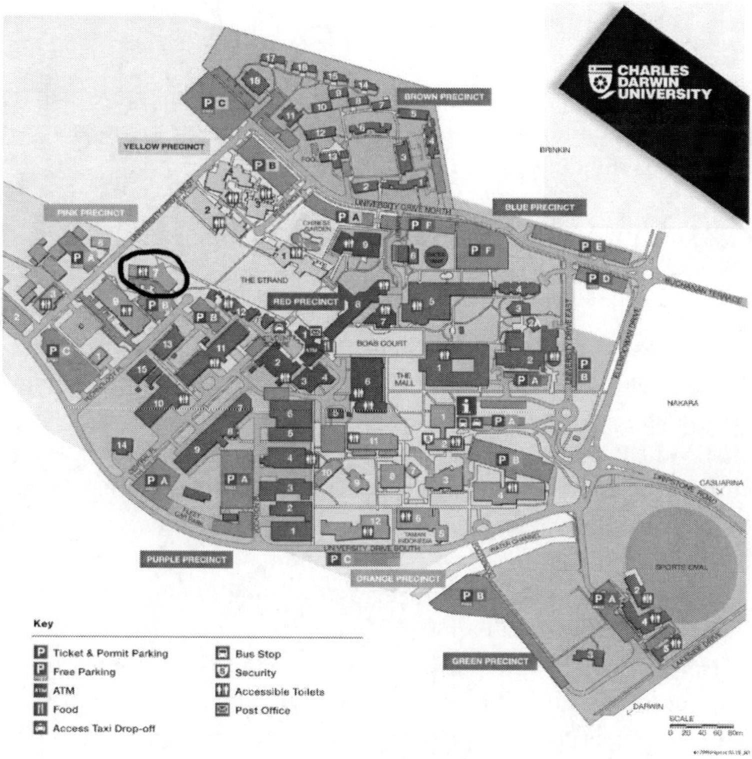

Figure 4. Charles Darwin University map and the selected building for investigation.

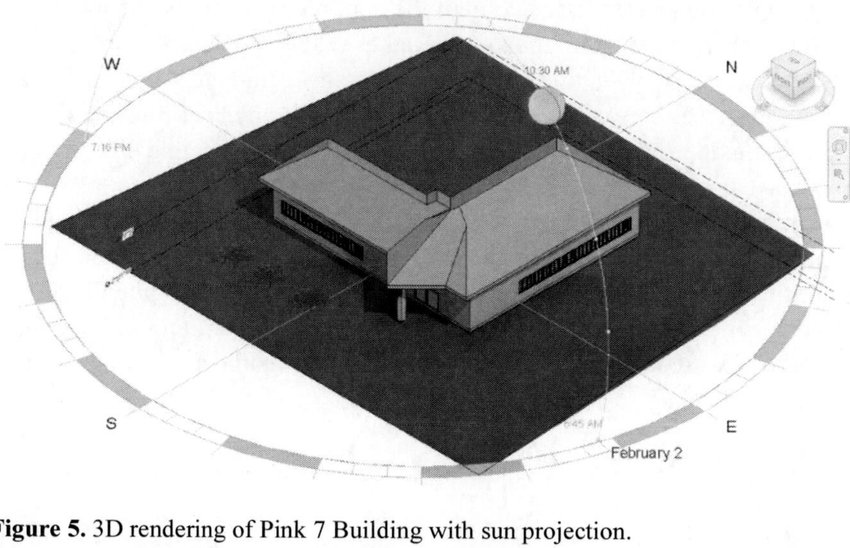

Figure 5. 3D rendering of Pink 7 Building with sun projection.

2.2. Load Calculation

For load calculation, zones were set up inside the building, in which a zone is an area inside the building which is controlled by a single thermostat. The Pink 7 Building was divided into two zones:

- Zone 1: consisting of all classrooms and offices
- Zone 2: consisting of laboratories and storeroom.

Using Revit software, three sets of cooling load were generated for comparison for three scenarios: (1) current construction; (2) using double glazed windows; (3) without windows.

The load calculations were carried out for the above three scenarios with the current construction being considered as the baseline. Observable reductions in cooling loads after glazing was compared with the baseline to select the most efficient window type for optimal energy consumption. Revit was used to calculate the cooling loads for all the window categories and equally used for the calculation of energy consumption and energy cost comparison for selection of best design options that would ensure optimal thermal comfort. A critical component of this software is the ASHRAE 90.1 standards which is used to predict both the cooling and heating loads of buildings. All the range of window-to-wall ratio for every region (including tropical regions) is contained in this standard. A number of assumptions were made to ensure an easy calculation of the cooling load using the software.

Other early studies (Mehdizadeh-Rad et al., 2022; Mempouo et al., 2010) have focused on evaluating the various glazing parameters such as glazing types and spaces between the glazing, so due to the practical considerations of the thermal performance and costs involved when selecting glazing, it is important to analyse three different scenarios including the current state of construction of the building, using double glazed windows and without any windows at all. This is important because the outcome would be the identification of economically efficient window condition despite the various glazing types with numerous numbers of panes that are commercially available for use during construction. The cooling load data is summarized in Table 1 below:

Table 1. Cooling load for three scenarios

	(1) Current	**(2) Double Glazing**	**(3) No Window**
Peak Cooling Total Load (kW)	44.9	45.0	42.4
Peak Cooling Month/ Hour	Jan 10am	Jan 10am	Nov 10am
Peak Cooling Sensible Load (kW)	23.3	23.3	22.0
Peak Cooling Latent Load (kW)	21.7	21.7	20.4
Maximum Cooling Capacity (kW)	44.2	44.8	41.2
Peak Cooling Airflow (CFM)	2,787	2,806	2,413
Peak Heating Load (Btu/h)	-40.284.6	-40,239.6	-41.654.0
Peak Heating Airflow (CFM)	1,233	1,233	1,233
Cooling Load Density (W/m^2)	138.07	138.38	130.42
Cooling Flow Density (CFM/SF)	0.80	0.80	0.69
Cooling Flow/ Load (CFM/ton)	218.32	219.29	200.10
Cooling Area/ Load (SF/ton)	274.17	273.55	290.25
Heating Load Density (W/m^2)	-36.31	-36.27	-37.54
Heating Flow Density (CFM/SF)	0.35	0.35	0.35

For the first scenario based on the current construction of the building, the peak cooling total load was 44.9kW with the peak cooling sensible load of 23.3kW and the peak cooling latent load of 21.7kW with the peak cooling airflow of 2,787CFM.

The second scenario using double glazed windows had similar cooling load of 45kW with the same peak cooling sensible load of 23.3kW and peak cooling latent load of 21.7kW. The peak cooling airflow was 2,806CFM, which was 19CFM higher than the first scenario.

For the third scenario without windows from all sides of the building, the peak cooling total load dropped down to 42.4kW with peak cooling sensible load of 22kW and peak cooling latent load of 20.4kW. The peak cooling airflow dropped to 2,413CFM which was a significant reduction compared with the first two scenarios. The hypothetical third scenario was run to check the limits for comparison.

Simulations involving different design options and other heat mitigation strategies for the reduction of cooling load and optimization of thermal comfort are covered in the next section.

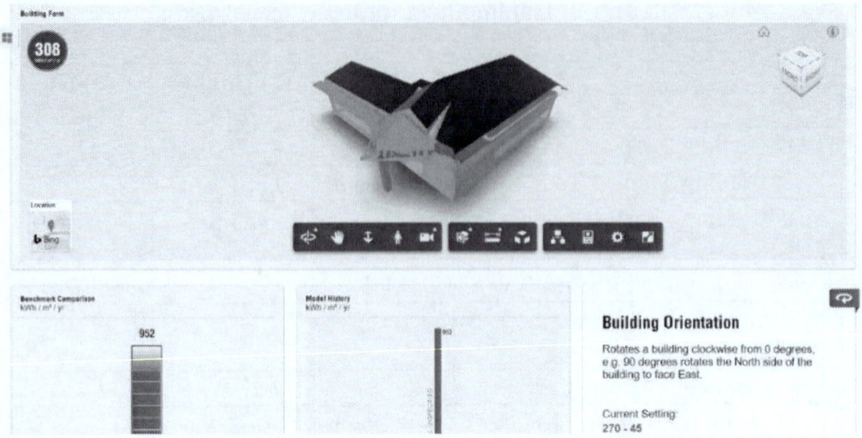

Figure 6. 3D Simulation by Revit Insight modelling.

2.3. Simulations

2.3.1. Insight Modelling

Insight is a plugin of Autodesk Revit 2021 that performs energy analysis and HVAC energy load calculation, which facilitates the design of more energy-efficient buildings (Figure 5). Autodesk Revit with Insight is basically used to analyze the energy performance of a building during the design stage. It has the advantage of optimizing the building design and guide engineers in making key decisions about the energy performance of a building (Carvalho et al., 2021; Gerrish et al., 2017)

For each simulation that must be run on Insight requires the generation of the energy model. After uploading the energy model, the official servers of Autodesk perform the calculations and send the results through e-mail. The simulations provide the energy consumption of the building and the energy cost incurred throughout the year.

2.3.2. Cool Coating

As mentioned earlier, cool coating is an effective technique to minimize the incoming solar radiation by reflecting most of the sunlight. A simulation was carried out after applying cool coating to the roof and walls using Revit software. A tabular representation of the total `energy consumption of the building 7 before and after cool coating is shown in (Table 2). From the table, there was a significant difference between the obtained results before cool coating was applied and afterwards. Although the heating flow density and peak heating airflow remained unchanged, the reduced cooling loads due to cool coating gave reduced energy demands and consumption by the model building as evidenced in the reduction in the values of the cooling load parameters after cool coating was applied.

2.3.3. Change of Windows

Windows are the weakest points regarding the thermal insulation of a building envelope. The Revit software gave an option to specify the size, position and orientation (window casement) of windows on the building's site. Various window sizes (24" × 36", 30" × 46", 28" × 46", 26" × 42", 24" × 42", 61" × 48", 36" × 72"…..) were available on the Revit interface. To obtain the best window for optimal and efficient energy consumption, various window orientations and sizes were applied as shown in (Figures 6 & 7). First, 24" × 36 dimension was applied, followed by 72" × 53" and the energy consumption of the building after the window size and orientation were varied as above was modelled using Revit. After changing the windows, the energy consumption of the building was reduced from 308kWh/m^2/year to 298kWh/m^2/year (Figures 6 & 7).

Table 2. Comparison before and after having cool coating

	(1) Before cool coating	**(2) After cool coating**
Peak Cooling Total Load (kW)	47.3	45.1
Peak Cooling Month/Hour	Jan 9am	Jan 10am
Peak Cooling Sensible Load (kW)	24.1	23.4
Peak Cooling Latent Load (kW)	23.3	21.7
Maximum Cooling Capacity (kW)	46.3	45.0
Peak Cooling Airflow (CFM)	2,951	2,839
Peak Heating Load (Btu/h)	-39,384.7	-40.104.8
Peak Heating Airflow (CFM)	1,233	1,233
Cooling Load Density (W/m^2)	145.57	138.73
Cooling Flow Density (CFM/SF)	0.84	0.81
Cooling Flow/Load (CFM/ton)	219.28	221.32
Cooling Area/Load (SF/ton)	260.05	272.87
Heating Load Density (W/m^2)	-35.50	-36.18
Heating Flow Density (CFM/SF)	0.35	0.35

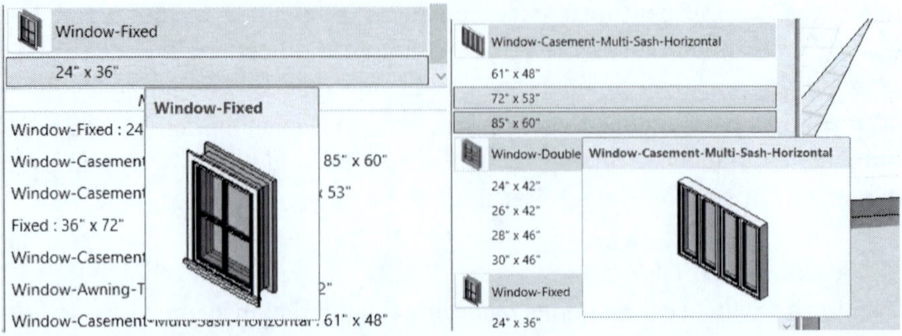

Figure 7. Change of windows (from left to right) for simulation.

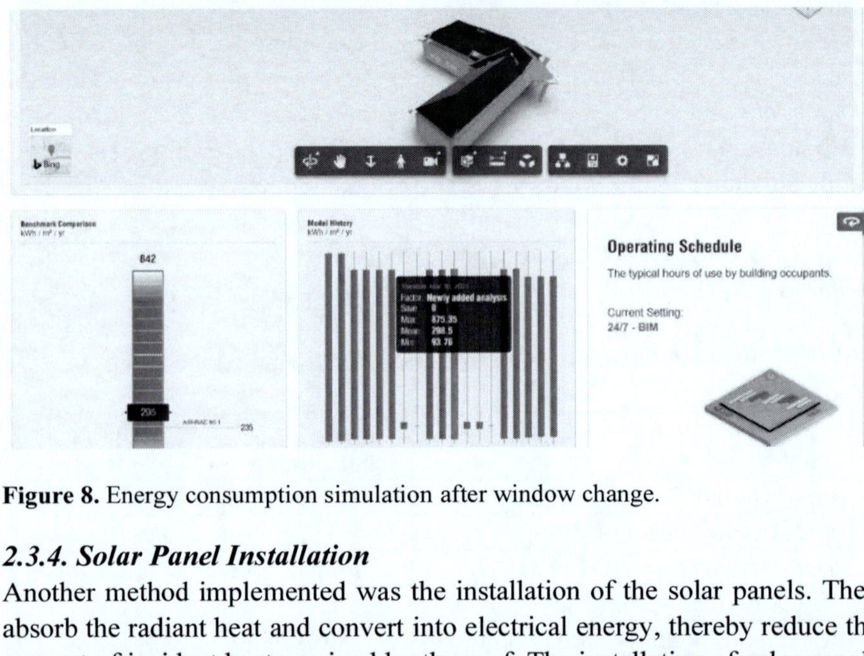

Figure 8. Energy consumption simulation after window change.

2.3.4. Solar Panel Installation

Another method implemented was the installation of the solar panels. They absorb the radiant heat and convert into electrical energy, thereby reduce the amount of incident heat received by the roof. The installation of solar panels further reduces the energy consumption of the building by minimizing the cooling and heating loads. The total area of the roof was 2996 m^2 and 589.54m^2 of the roof area was installed with solar panel for simulation. As this experiment was purely simulation, Revit computed the total electricity generated without displaying it and according to the simulation result the energy consumption reduced from 298kWh/m^2/year to 295kWh/m^2/year.

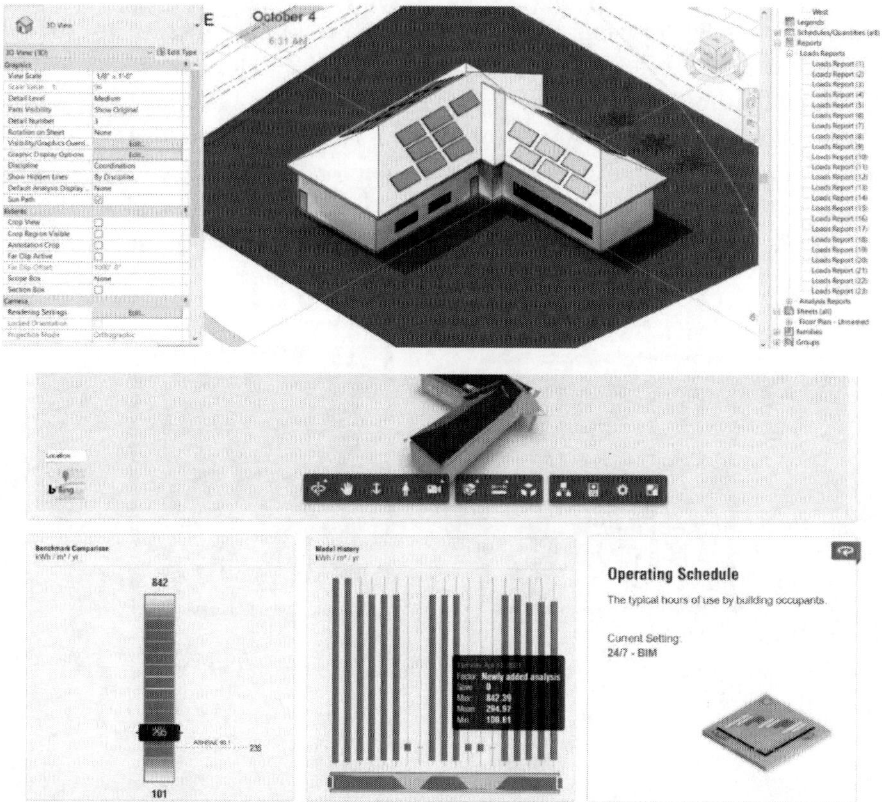

Figure 9. Solar panel installation and energy consumption simulation afterwards.

3. Results and Discussion

In this chapter, the reduction of cooling load of the Pink 7 Building in Darwin, Australia was investigated using both Revit software and EnergyPlus software for determination of desired thermal comfort. The results from this study shown below has given more insight of the thermal comfort and cooling load.

3.1. Simulation Results

Detailed annual energy consumption report and detailed annual simulation report generated by Revit software are accessible via the following QR codes (Figure 9).

After modelling using Revit software and testing in EnergyPlus, the pie chart generated by Revit simulations in Figure 10 depicts the energy consumption for the whole year by cooling and lighting and equipment. It is clear that most of the energy is consumed for cooling. Therefore, the energy expenditure for acquiring the thermal comfort using air conditioners is comparatively high. High energy consumption leads to high carbon dioxide emission which has to be mitigated to address the pressing issue of global warming.

Figure 10. QR codes for Detailed Annual Simulation Report (left) and detailed Annual Energy Consumption Report (right).

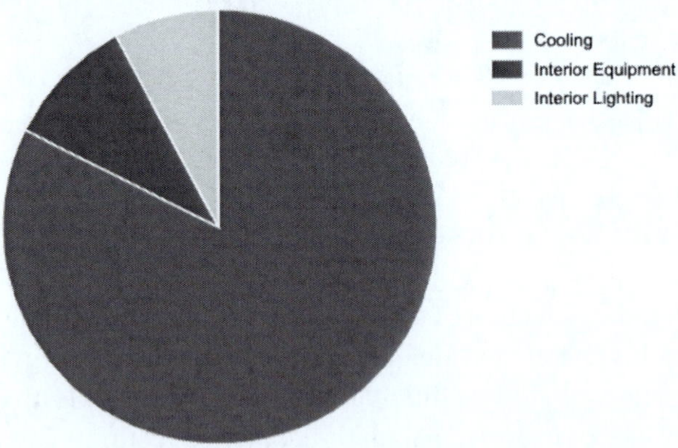

Figure 11. Energy consumption chart.

Figure 11 shows the comparison of the energy consumption results obtained for various simulations in this study, ranging from original construction, change of windows, cool coating, double-glazing to solar panel installation.

The grey bar of the graph represents the mean energy consumption of the building as per its normal construction parameters, which is 308kWh/m^2/year. The black bar represents the energy consumed by the building after changing the window configuration. This change has helped to bring down the mean value of energy consumption by 1kWh/m^2/year to 307kWh/m^2/year. The blue bar represents the mean energy consumption of the building after the treatment of cool coating on the roof and walls. This method proved to be highly efficient as it significantly reduced the energy consumption from 307kWh/m^2/year to 300kWh/m^2/year, thus providing with the energy reduction of 7kWh/m^2/year. The orange bar represents the mean energy consumption obtained after the building has double-glazing windows with a 2kWh/m^2/year of energy reduction. The purple bar depicts the mean energy consumption of the building after solar panel installation. The installation of solar panels is costly as compared to other methods of reduction. The expenditure includes the material cost of solar panels, the installation cost, the setup of the system by an authorised team, permissions from energy providers and authorities and long-term maintenance cost. Although the initial cost of solar panel installation is significant, it is recoverable in the long run. Using solar panels reduce the mean energy consumption from 298kWh/m^2/year to 295kWh/m^2/year, thereby saving 3kWh/m^2/year of energy. Additional solar panel installation yields more reduced energy consumption of a building, consequently, a second set of solar panels were installed. The final bar on the graph i.e., the red bar represents the mean energy consumption after the installation of the second set of solar panels. As such, the energy consumption is reduced from 295kWh/m^2/year to 293kWh/m^2/year, providing with energy saving of 2kWh/m^2/year.

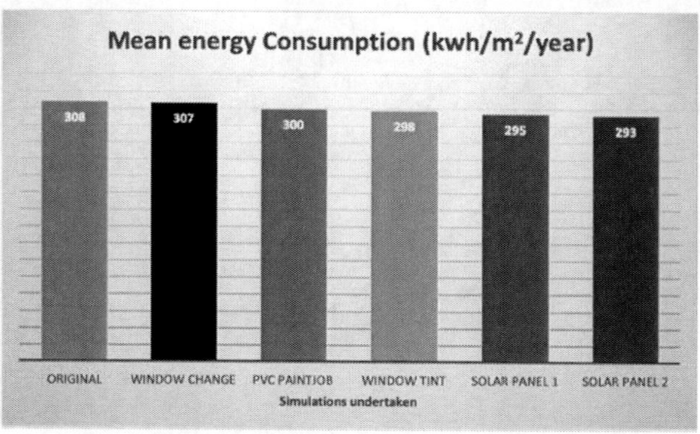

Figure 12. Energy consumption comparison for various strategies.

3.2. Effectiveness

The effectiveness of each strategy used is calculated by comparing its ability to minimize the mean energy consumption to the total mean energy consumption. The pie chart in Figure 12 compares the energy effectiveness of each strategy. Changing the windows was proved to be the least effective for reducing the cooling load, contributing to just 7% of the total effectiveness. The second least effective technique was double-glazing of windows which provided with the energy reduction of 2kWh/m^2/year which contributes to 13% effectiveness. The second most effective methods used in this research was the installation of solar panels, which provided with the total energy saving of 5kWh/m^2/year, contributing to 33% effectiveness of the operation.

Since the roof and wall are very critical parts of the building envelope and plays very key roles as energy barrier systems, the best strategy in this research was the cool coating on the roof and walls, contributing to a total of 47% (7kWh/m^2/year) of the energy effectiveness and cooling load reduction through overall solar reflectance.

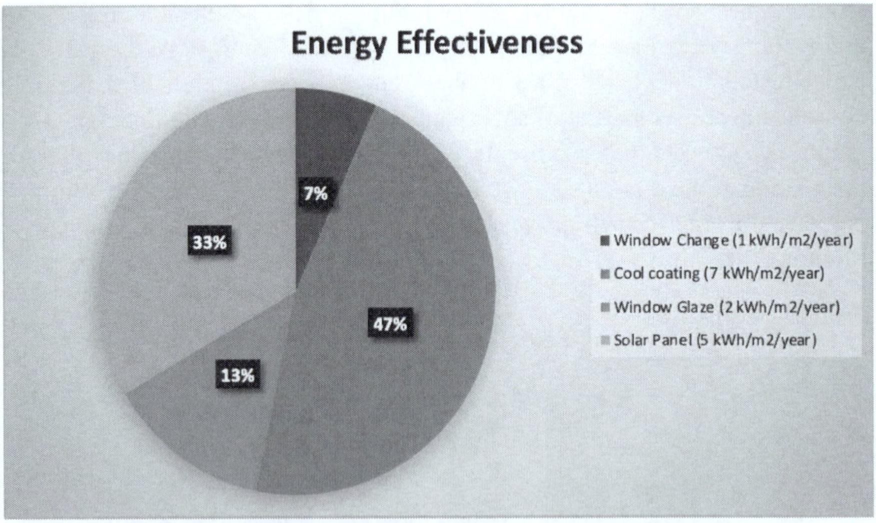

Figure 13. Effectiveness of each strategy.

3.3. Cost

Another factor to be considered is the cost involved in these strategies regarding the installation, operation and long-term maintenance, which are elaborated below:

3.3.1. Change of Windows

A total of 21 windows of the building were involved. After the change of windows, the running cost was reduced from A$80.08/m^2/year to A$79.82/m^2/year. The average material cost and installation of each window in 2021 was A$450, so the total cost of changing all the windows was A$9,450.

3.3.2. Window Glazing

The process of converting existing windows to become double-glazing is costly because it requires disassembly of the unit and then re-assembling it again for double glazing. The running cost was reduced from A$80.08/m^2/year to A$79.56/m^2/year after double glazing the windows. The total installation cost involved was estimated to be A$28,350 in 2021.

3.3.3. Solar Panel

As compared to the original running cost of the building of A$80.08/m^2/year, the new running cost after solar panel installation came out to be $78.78/m^2/year. Various factors that affect the installation cost are: the location of the building, the size of panels, on-grid or off-grid, solar rebates, the orientation of the building and so on. According to the data from Solar Choice' installer network database, a fully installed 20kW solar panel system costed approximately A$15,000-A$20,000 in 2021 (Choice, 2021), so it is economically feasible to run on solar panels.

3.3.4. Cool Coating

This has proved to be the most effective technique in minimizing the cooling load of the building. If compared to the original cost of A$80.08/m^2/year, the new cost after having cool coating on the roof and walls was reduced to be A$78.52/m^2/year. The total area to be covered as per the simulation was 589.54m^2 (roof area) and 448.94m^2 (area of the walls). The average cost of material was $26.75/m^2 in 2021 (excluding the labour charges). The total cost, including that of labour for having cool coating would be around A$30,000.

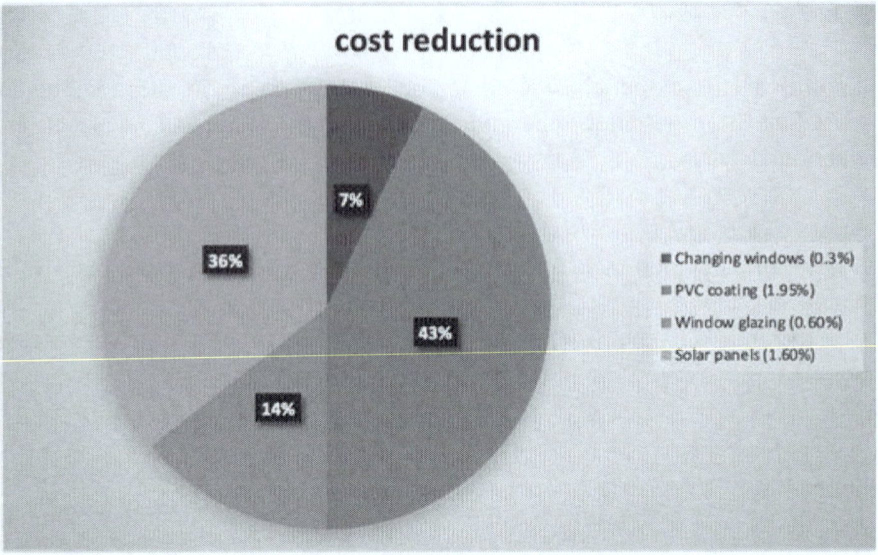

Figure 14. Cost Effectiveness of the reduction methods.

The pie chart in Figure 13 shows the cost effectiveness of each strategy. Cool coating was proved to be the most effective method as it brings down the cost by A\$1.56/m^2/year of the whole project. Solar panel Installation was the second most effective technique used in this research work as it reduced the cost by A\$1.3/m^2/year. Double glazing/ tinting was another method which reduced the cost by A\$0.52/m^2/year and last but not the least, change of windows reduced the cost by A\$0.26/m^2/year.

However, there are few issues worth mentioning here:

1. The average life expectancy of cool paint is 10 years (Zinzi & Fasano, 2009) as per the industrial reports and standards but considering extreme weather conditions of Darwin just to be on the safe side, consider repainting after 8 years just for maximum effectiveness of the cool coatings.
2. On another note, the average energy generation of the solar panels is 20kWh (Roy et al., 2016) depending upon the weather conditions. If all the 30 components installed at the location function at the full capacity 600kWh/day can be easily generated which is a surplus amount of energy generation. The cooling load of the building is 45 kW and if other electrical equipment is run as well, the total energy consumption can be expected around 100 kWh/day. Now that we

have surplus of 500 kWh/day, it can be supplied back to the power grid for which the energy company pays 0.83 cents/kW (Solar quotes, 2009), which could generate a whopping 415 AUD/day. Thus, if the solar panels run at full efficiency for at least 1.5 years, the whole amount invested into the project can be regained.

Conclusion

In conclusion, based on the research conducted in this chapter, the cooling load can be reduced using various strategies while maintaining the thermal comfort. The comparison from the simulations is valuable to understand the cost-effectivess of each strategy. Less costly methods such as changing the design of windows and glazing or tinting of windows were quite effective. Cool coating must be re-applied after a period of seven to eight years. On the other hand, installation of solar panels proved to be a better investment in the long run. Although the initial set-up costs are high, they are recoverable in a shorter period as per the simulation results and solar rebates available. These are the strategies that can reduced the energy consumption for cooling and peak loads, which can contribute to mitigate the urban heat island effect.

Besides the strategies mentioned in the simulations, other methods for reducing the cooling loads of buildings such as: airtightness of the building envelope, building insulation, shading options, green walls and roofs are worthy for further investigation.

References

ASHRAE. (2012). *American Society of Heating, Refrigerating and Air-Conditioning Engineers Handbook*. ASHRAE.

Bhatia, A. (2004). *Heat Loss Calculations and Principles*. *877*.

Carvalho, J., Almeida, M., Bragança, L., & Mateus, R. (2021). BIM-Based Energy Analysis and Sustainability Assessment—Application to Portuguese Buildings. *Buildings*, *11*(6), 246. https://doi.org/10.3390/buildings11060246

Chai, Q., & Zhang, X. (2010). Technologies and policies for the transition to a sustainable energy system in china. *Energy*, *35*(10), 3995–4002. https://doi.org/10.1016/j.energy.2010.04.033

Choice, S. (2021). *20kW commercial solar power systems: Pricing, output, and returns*. https://www.solarchoice.net.au/blog/20kw-commercial-solar-power-systems-pricing-output-and-returns/#:~:text=How much does a 20kW,%2420%2C000 as of November 2021

Darwin, C. of. (2021). *Feeling Cooler in Darwin: Darwin heat mitigation and adaptation strategy*. https://research.csiro.au/darwinlivinglab/wp-content/uploads/sites/278/2021/12/Darwin-Heat-Mitigation-and-Adaptation-Strategy.pdf

Evans, S., Marchand, R., & Ackerman, T. (2014). Variability of the Australian Monsoon and Precipitation Trends at Darwin. *Journal of Climate*, *27*(22), 8487–8500. https://doi.org/10.1175/JCLI-D-13-00422.1

Gerrish, T., Ruikar, K., Cook, M., Johnson, M., Phillip, M., & Lowry, C. (2017). BIM application to building energy performance visualisation and management: Challenges and potential. *Energy and Buildings*, *144*, 218–228. https://doi.org/10.1016/j.enbuild.2017.03.032

Lepp, N. W. (2008). Planting Green Roofs and Living Walls. *Journal of Environmental Quality*, *37*(6), 2408–2408. https://doi.org/10.2134/jeq2008.0016br

Luksamijarulkul, P., & Pipitsangjan, S. (2015). Microbial Air Quality and Bacterial Surface Contamination in Ambulances During Patient Services. *Oman Medical Journal*, *30*(2), 104–110. https://doi.org/10.5001/omj.2015.23

Mamtora, J. (2011). Thinking Big Picture: Meeting the Needs of Researchers in Northern Australia. *Australian Academic & Research Libraries*, *42*(2), 88–102. https://doi.org/10.1080/00048623.2011.10722216

Mattheos, S., Shamila, H., Giulia, U., Jonathan, F., Riccardo, P., Afroditi, S., Francesco, F., & Samira, G. (2017). *Heat Mitigation Program Darwin NT: Final Report for the Northern Territory Government*. https://apo.org.au/sites/default/files/resource-files/2017-09/apo-nid183291.pdf

McQuiston, F. C. (2004). *Heating, ventilation and air conditioning* (6th ed.). Wiley.

Mehdizadeh-Rad, H., Choudhry, T. A., Ng, A. W. M., Rajabi, Z., Rais, M. F., Zia, A., & Tariq, M. A. U. R. (2022). An Energy Performance Evaluation of Commercially Available Window Glazing in Darwin's Tropical Climate. *Sustainability*, *14*(4), 2394. https://doi.org/10.3390/su14042394

Mempouo, B., Cooper, E., & Riffat, S. B. (2010). Novel window technologies and the Code for Sustainable Homes in the UK. *International Journal of Low-Carbon Technologies*, *5*(4), 167–174. https://doi.org/10.1093/ijlct/ctq013

Meteorology, B. of. (2022). *Climate statistics for Australian locations*. http://www.bom.gov.au/climate/averages/tables/cw_014015_All.shtml

Nithiarasu, P., Lewis, R. W., & Seetharamu, K. N. (2016). *Fundamentals of the finite element method for heat and mass transfer*. John Wiley & Sons.

Omer, A. (2002). Low energy building materials: an overview. In *Advances in Building Technology* (pp. 1257–1263).

Omer, A. M. (2008). Renewable building energy systems and passive human comfort solutions. *Renewable and Sustainable Energy Reviews*, *12*(6), 1562–1587.

Pérez-Lombard, L., Ortiz, J., & Pout, C. (2008). A review on buildings energy consumption information. *Energy and Buildings*, *40*(3), 394–398. https://doi.org/10.1016/j.enbuild.2007.03.007

Prado, R. T. A., & Ferreira, F. L. (2005). Measurement of albedo and analysis of its influence the surface temperature of building roof materials. *Energy and Buildings*, *37*(4), 295–300. https://doi.org/10.1016/j.enbuild.2004.03.009

Quotes, S. (2009). *NT Feed in Tariff Information*. Solar Systems. https://www.solarquotes.com.au/systems/feed-in-tariffs/nt/

Rizwan, A. M., Dennis, L. Y. C., & Liu, C. (2008). A review on the generation, determination and mitigation of Urban Heat Island. *Journal of Environmental Sciences*, *20*(1), 120–128. https://doi.org/10.1016/S1001-0742(08)60019-4

Roy, S., Malkhandi, A., & Ghose, T. (2016). Modeling of 5kW single phase grid tied solar photovoltaic system. *2016 International Conference on Computer, Electrical & Communication Engineering (ICCECE)*, 1–8. https://doi.org/10.1109/ICCECE.2016.8009539

Sait, H. H. (2013). Estimated Thermal Load and Selecting of Suitable Air-conditioning Systems for a Three Story Educational Building. *Procedia Computer Science*, *19*, 636–645. https://doi.org/10.1016/j.procs.2013.06.085

Santamouris, M., Synnefa, A., & Karlessi, T. (2011). Using advanced cool materials in the urban built environment to mitigate heat islands and improve thermal comfort conditions. *Solar Energy*, *85*(12), 3085–3102. https://doi.org/10.1016/j.solener.2010.12.023

Sayigh, A., & Marafia, A. H. (1998). Thermal comfort and the development of bioclimatic concept in building design. *Renewable and Sustainable Energy Reviews*, *2*(1–2), 3–24.

Taleghani, M., Tenpierik, M., Kurvers, S., & van den Dobbelsteen, A. (2013). A review into thermal comfort in buildings. *Renewable and Sustainable Energy Reviews*, *26*, 201–215. https://doi.org/10.1016/j.rser.2013.05.050

Yang, L., Yan, H., & Lam, J. C. (2014). Thermal comfort and building energy consumption implications – A review. *Applied Energy*, *115*, 164–173. https://doi.org/10.1016/j.apenergy.2013.10.062

Zinzi, M., & Fasano, G. (2009). Properties and performance of advanced reflective paints to reduce the cooling loads in buildings and mitigate the heat island effect in urban areas. *International Journal of Sustainable Energy*, *28*(1–3), 123–139. https://doi.org/10.1080/14786450802453314

About the Editors

Prof Satyaprakash

Dr Satyaprakash is a researcher in the field of Geo-Informatics, besides Building Materials. His active research projects include use of satellite data for disaster mitigation, water quality studies, besides Land Surface Temperature/ Urban Heat Island. For the last 30 years in this field, he has done research in the field of geodetic GPS mapping over Himalayas for mapping the crustal deformation studies, use of GIS for utility mapping, InSAR data for surface deformation mapping and use of Google Earth Engine for studying the dynamics of earth. He is also interested in processing of Ecostress data for climate change studies and Hyperspectral data for mineral and mining exploration. He has filed 25 innovative and design patents and has more than 50 publications in international journals and is on the reviewer panel of several of the international journals.

Dr Anne W.M.Ng

Dr Ng's PhD thesis entitled "parameter optimisation of river water quality models using genetic algorithms" was awarded Victoria University Vice Chancellor's Citation for Excellence in Research in 2003. She was the first researcher to develop a Yarra River model that mimics the river condition and to study natural treatment in river systems. Her field of expertise is to contribute to the body of knowledge in the applied area of sustainable water resources. This contribution can then be carried out in conjunction with industry, government and the community with emphasis on research in "need". Dr Ng's field of expertise evolve from the general areas of surface water hydrology, and natural resource management. Over her 20 years of academic career, she has also explored the use of artificial Intelligence to enhance and to optimise the overall modelling processes in the field of drought and flood management, and stormwater pipe deterioration in preventing flooding. Integrated Urban Water Management (IUWM). Is another branch within water resources to address issues mainly on "sustainability in water resources management. IUWM is the most important research of the present and into the future due to climate change and urbanisation. Dr Ng has build a strong track record in this arena including green infrastructure and the use to mitigate Urban Heat Island effect. She is also the lead/co-lead Guest Editor in three special edition of the Water Journal MPDI (Impact factor of 2.544); *"Urban Green Infrastructure: Current Issues, Challenges and Opportunities*", "*Innovate Approaches to Sustainable water resources management under population growth, lifestyle improvements and climate change*" and "*Addressing water security issues through water recycling, harvesting footprints, virtual water trade, and other novel approaches*".

List of Contributors

G. S. N. V. K. S. N. Swamy
Jagabandhu Panda
Asmita Mukherjee
Sudhansu S. Rath
Raphael Chukwuka Nnachi
Anne W. M. Ng
Muhammad Atiq Ur Rehman Tariq
Satyaprakash
Pinakana Sai Deepak
Vaibhav Rai Khare
Priyanka Kumari
Vaishaly
Mohit Saini
Hing-Wah Chau
Elmira Jamei
Saeed Kimiaei
Mostafizur Rahman
Hooman Mehdizadeh-Rad

Index

A

B

F

Q

R

S

T

U

V

W

Y

Z